Process Development Symposium 2022

Philadelphia, Pennsylvania, USA
8 – 9 June 2022

ISBN: 978-1-7138-6319-9

Printed from e-media with permission by:

Curran Associates, Inc.
57 Morehouse Lane
Red Hook, NY 12571

Some format issues inherent in the e-media version may also appear in this print version.

Printed with permission by Curran Associates, Inc. (2023)

For permission requests, please contact AIChE
at the address below.

AIChE
120 Wall Street, FL 23
New York, NY 10005-4020

Phone: (800) 242-4363
Fax: (203) 775-5177

www.aiche.org

Additional copies of this publication are available from:

Curran Associates, Inc.
57 Morehouse Lane
Red Hook, NY 12571 USA
Phone: 845-758-0400
Fax: 845-758-2633
Email: curran@proceedings.com
Web: www.proceedings.com

Process Development Symposium 2022

Philadelphia, Pennsylvania, USA
8 – 9 June 2022

AIChE
The Global Home of Chemical Engineers

TABLE OF CONTENTS

EMERGING TECHNOLOGIES

BATCH VS. CONTINUOUS, SCALE-UP & SCALE-DOWN, CASE STUDIES

INNOVATIONS IN PROCESS SAFETY

ONLINE HPLC FOR CONTINUOUS REACTION MONITORING IN CGMP MANUFACTURING

AICHE PROC DEV SYMPOSIUM JUNE 8, 2022

Martin Johnson, Joel Calvin, Bradley M. Campbell, Kevin Cole, Kenneth Desmond, Paul Desmond, Olivia Gowran, Jennifer Groh, Philip C. Hoffman, Declan Hurley, Gordon R. Lambertus, Carla Luciani, Scott May, Rob Moylan, Christopher S. Polster, Jeff Roberts, Richard D. Spencer, Luke Webster

Lilly

High Pressure Gas/Liquid reaction in vertical pipes in series flow reactor

Reductive Amination, 2000 kg

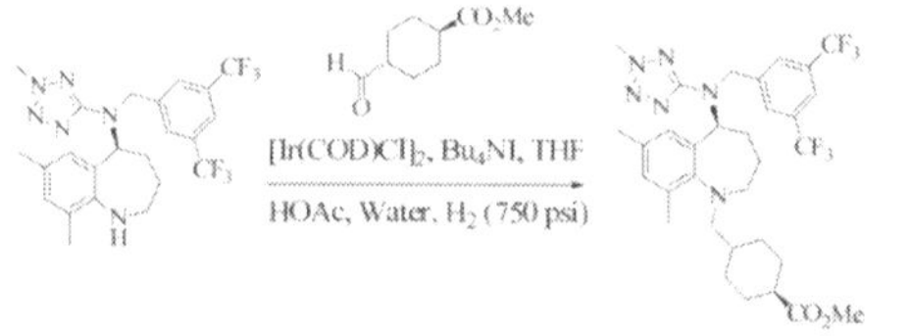

No existing reactor batch or flow

Safer Operation

"low risk operation"

Lilly

Org. Process Res. Dev., 2016, 20 (11), pp 1870–1898

Workup batch

Continuous reductive amination in GMP manufacturing

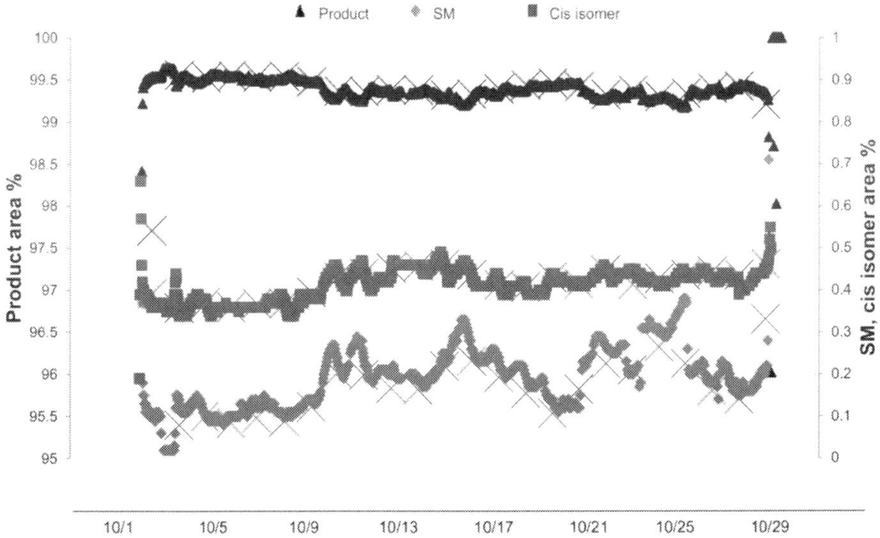

Org. Process Res. Dev., 2016, 20 (11), pp 1870–1898

Kilogram-scale prexasertib monolactate monohydrate synthesis

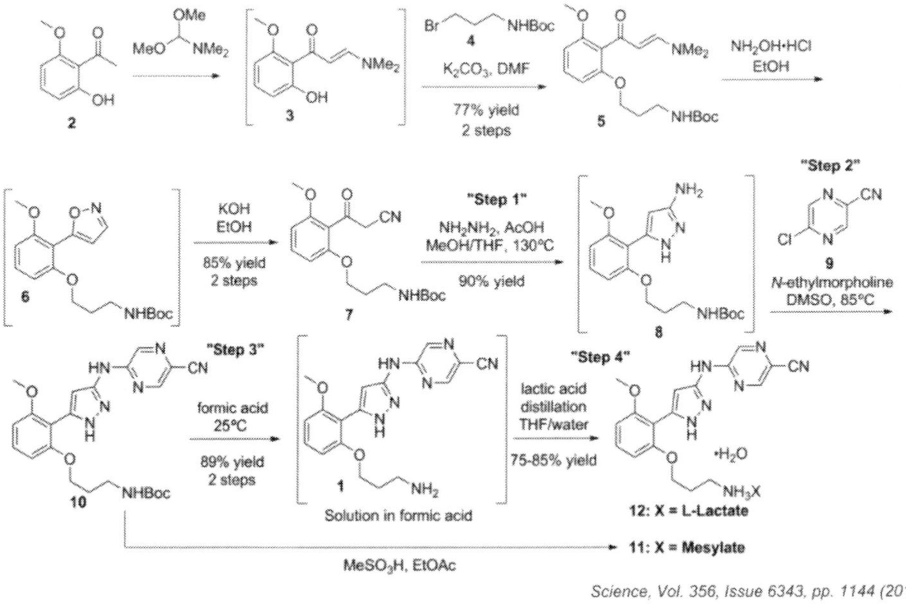

Science, Vol. 356, Issue 6343, pp. 1144 (2017)

2014-2015 GMP SVC Campaign

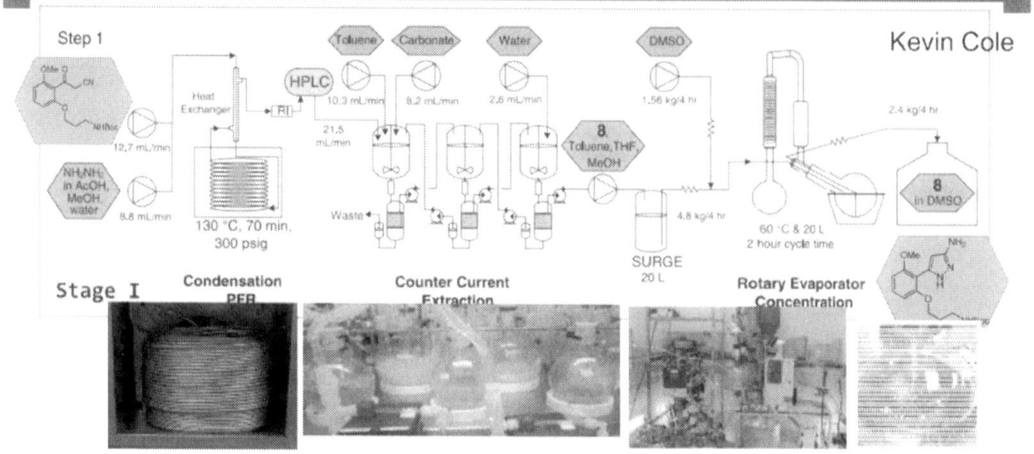

- SS PFR, 130 °C, 500 psi, reduced hydrazine equivalents compared to batch, better safety for hydrazine reaction, only 20 g NH2NH2 in PFR.
- Continuous counter-current extraction controlled residual Hydrazine from 6000 ppm to <2 ppm, acetic acid, deprotected pyrazole. Avoided isolation.
- Assay yields of 97–99%

Science, Vol. 356, Issue 6343, pp. 1144 (2017)

2014-2015 GMP SVC Campaign

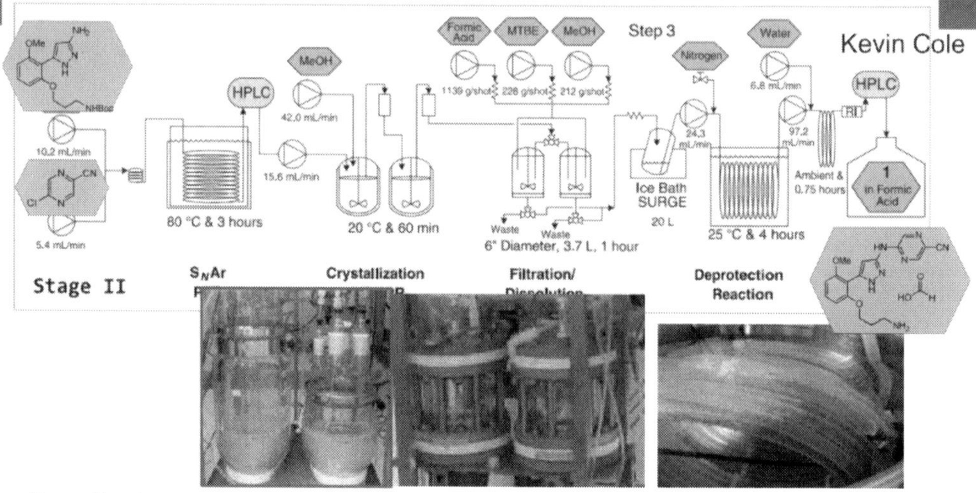

- Crystallization was important to impurity control strategy.. pyrazine and regio isomers. Purity of the crystallized solids was in excess of 99.8 area%
- "In-situ isolation". No handling of genotoxic, cytotoxic intermediate with OEL = 1 µg/m3
- Gas-Liquid PFR used to allow for gas sweeping & volatile removal

Science, Vol. 356, Issue 6343, pp. 1144 (2017)

Online HPLC for hydrazine condensation and Nucleophilic substitution reactions in flow

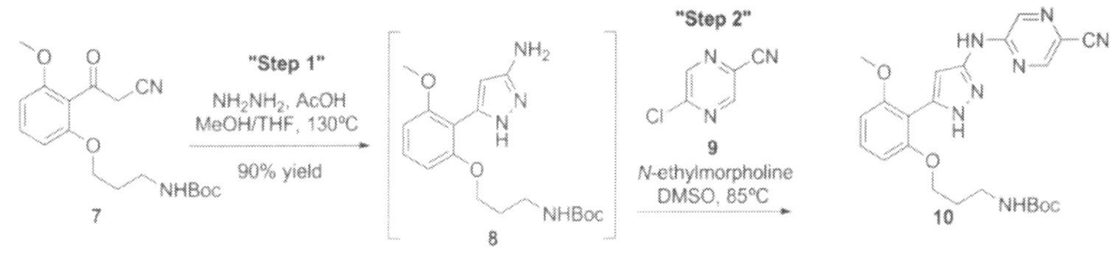

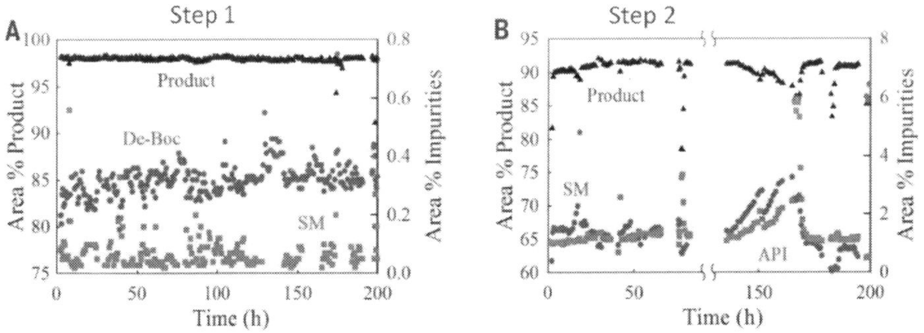

Science, Vol. 356, Issue 6343, pp. 1144 (2017)

7

New Facility for Fully Continuous Drug Substance Processes

Investment in Lilly Manufacturing, Kinsale Ireland
Facility completed Q3 2017
First project in new facility Q4 2017-Q1 2018

Lilly

Lilly Small-Volume CM Facility

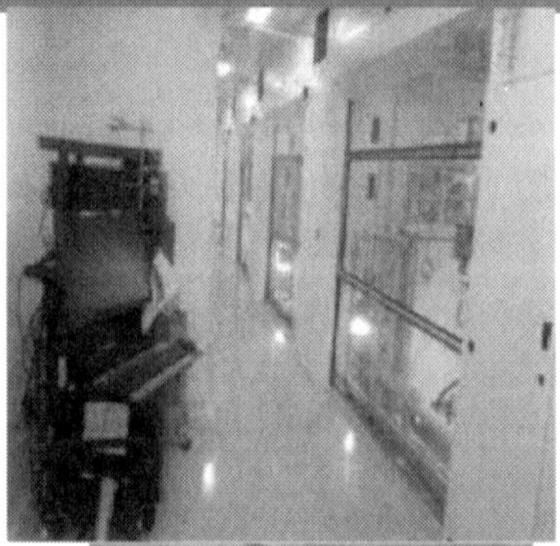

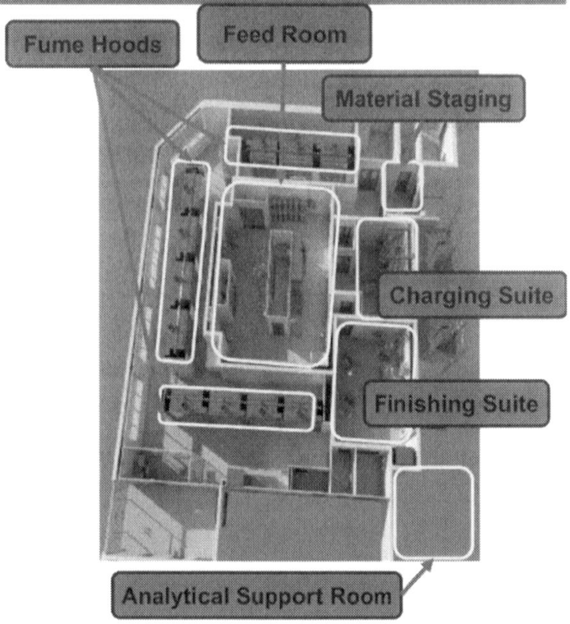

Sampling and Dilution skid

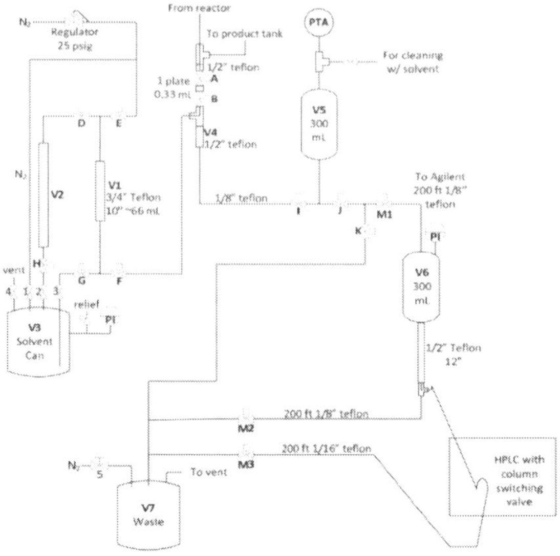

Org. Process Res. Dev., **2016**, 20 (5), pp 888–900, Figure S25

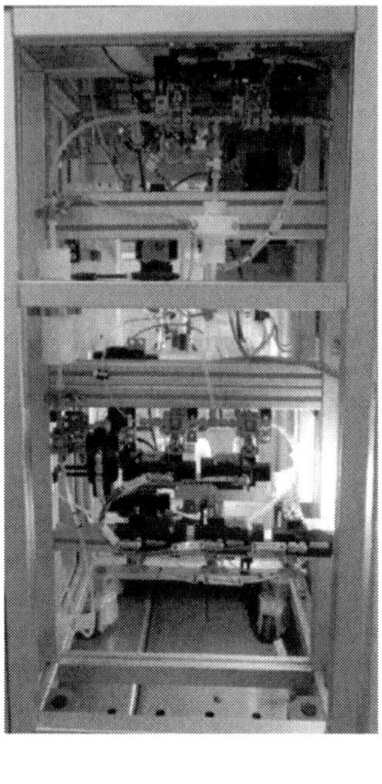

Online HPLC Data Acquisition and Transfer of Results

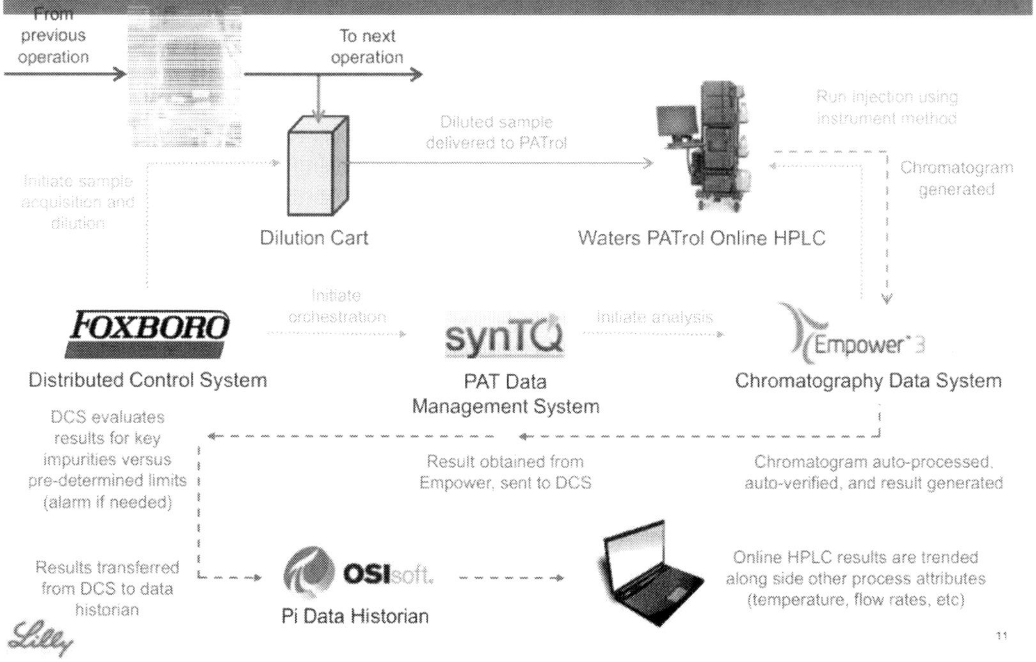

Chemistry – First Process in New Facility

Joel Calvin

Flow Process

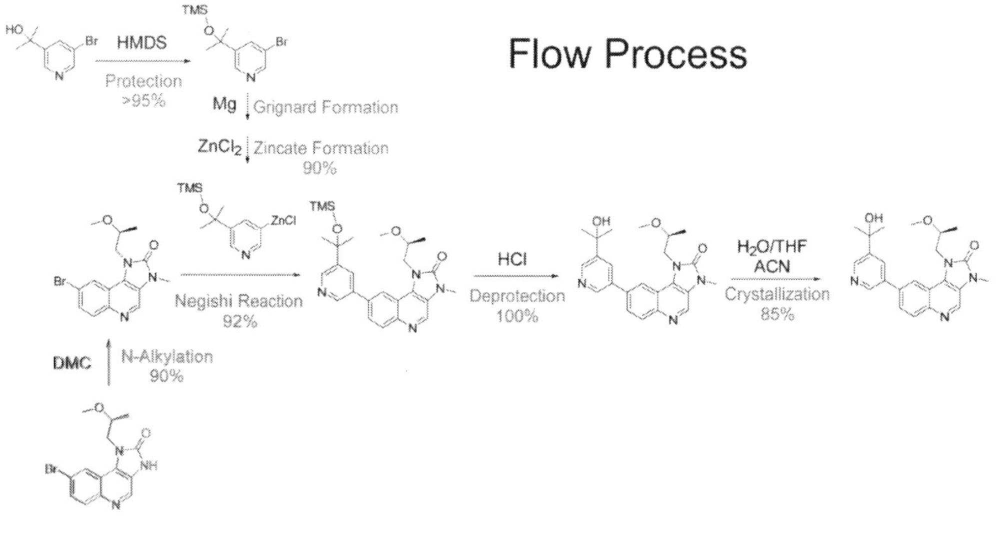

12

High-level schematic of the continuous process flow in cGMP manufacturing

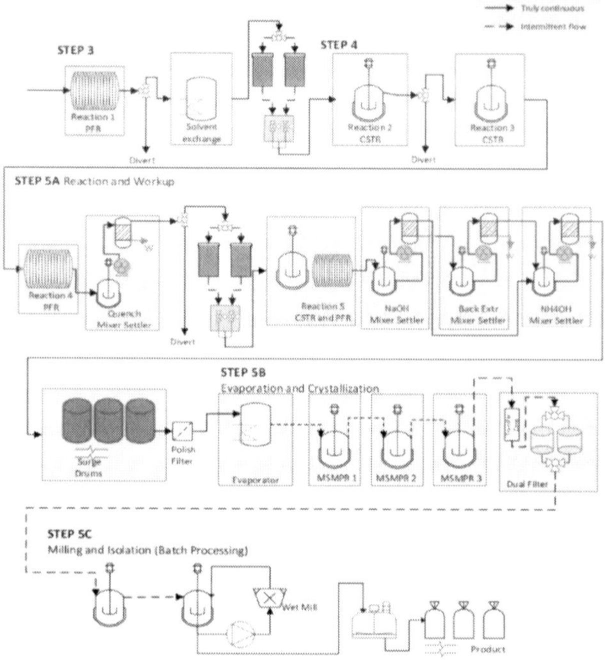

Lilly

13

Continuous Unit Operations & Residence Times

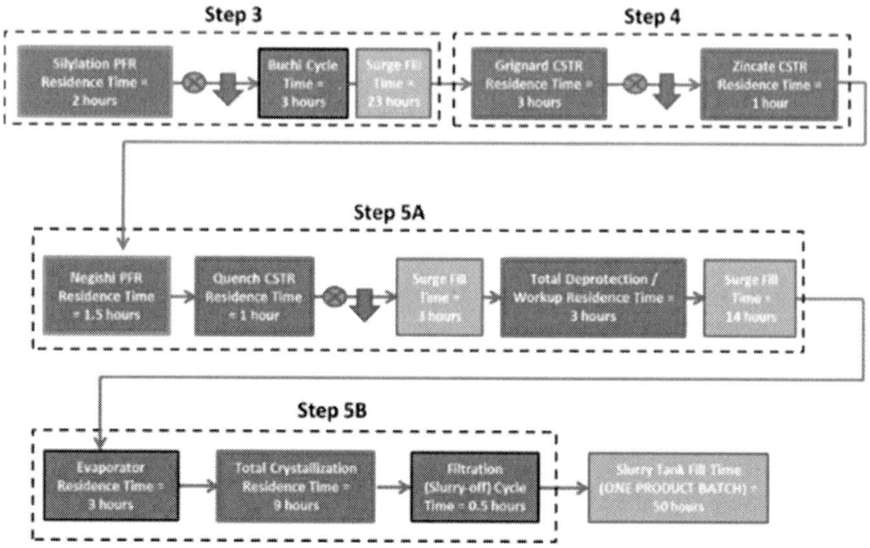

Contrast with drug product CM residence time and RTD, analytical frequency.

Lilly

On-line HPLC for silylation reaction

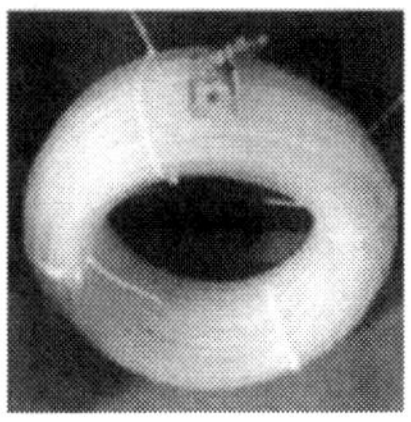

Silylation PFR
9.9 L PFA tube,
0.75" od 0.625" id
50 m, 120 minute τ

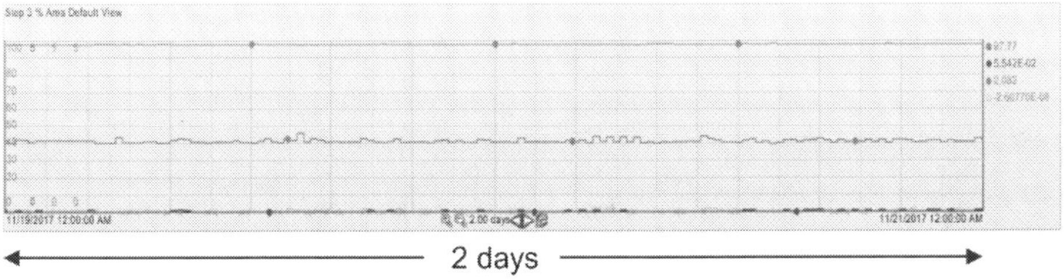

Green = product 97.8%, Blue = unreacted starting material 0.05%

Lilly

On-line HPLC for Grignard formation reaction

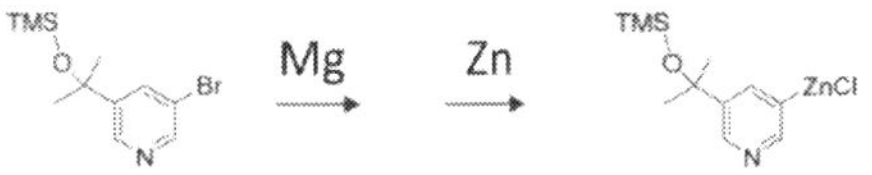

Grignard formation 10 L glass
3 hour τ

Zincate Formation
5 L glass
31 minute τ

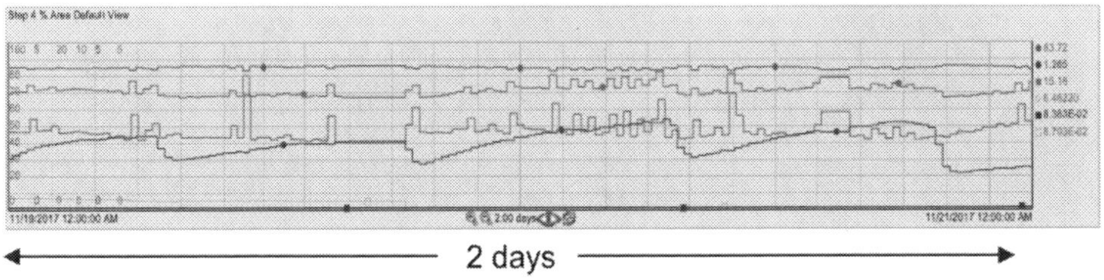

Green = product 84%, Blue = unreacted starting material 1% to 3% oscillating

Lilly

On-line HPLC for Negishi coupling reaction

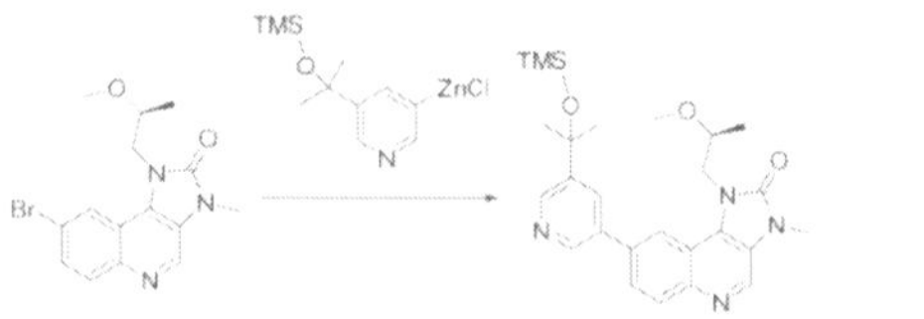

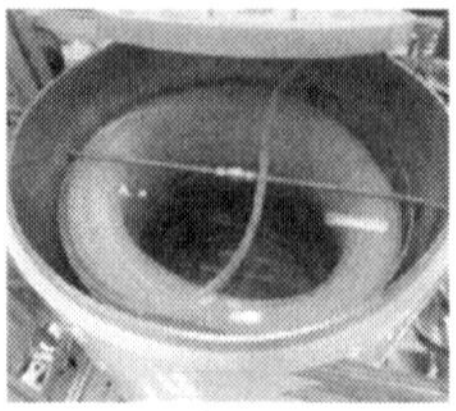

24 L PFA tube,
0.5" od 0.376" id
330 m, 75 minute τ

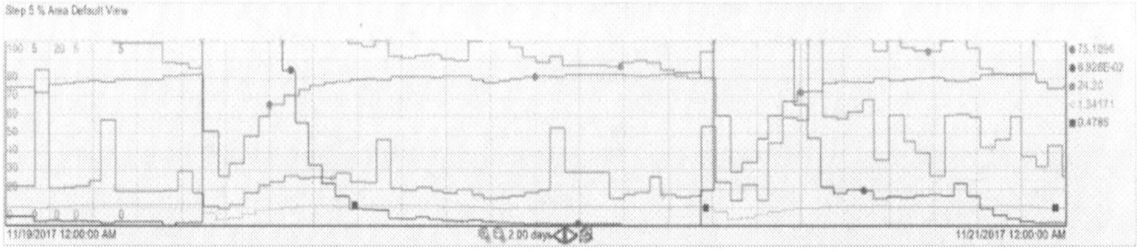

2 days

Green = product >80% typically, Blue = unreacted starting material <0.1% typically

Delay coil

gPROMS Model

Carla Luciani

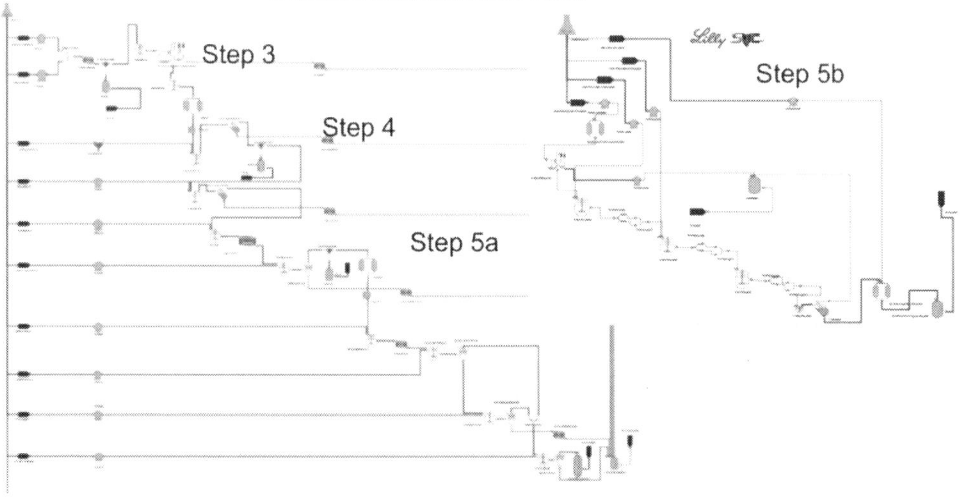

Response to 1 h tracer pulse at 10%

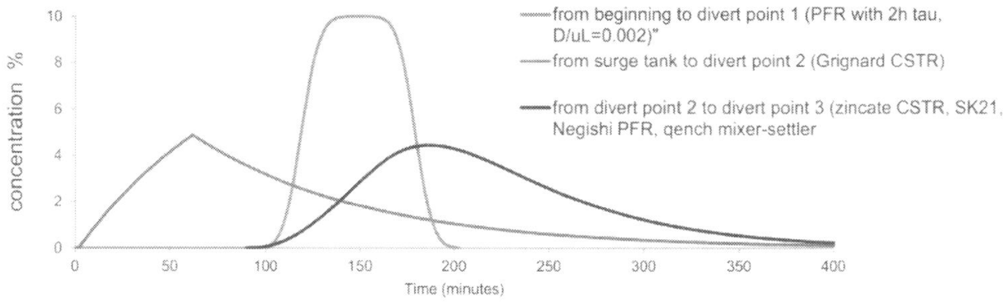

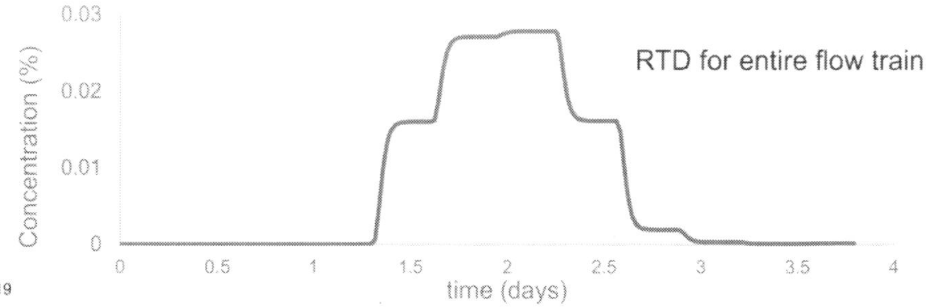

19

Online HPLC Conclusions

- monitored unreacted starting material and impurities at <0.1% level
- revealed insights as to how slight temperature changes impacted reaction conversion
- detected disturbances that could not be detected from other process parameters
- revealed insights as to how Mg charges impacted conversion
- used for divert and return from divert decisions and process adjustments
- Enabled by custom sampling/dilution/mixing/parking skids
- Data integrated by SynTQ
- Numerical modeling proved that 1h PAT frequency was sufficient to control the process.

Lilly

RAPID® Transforming Process Industries

SAPPHIRE – Accelerating Modular Process Development (aka the Corundum Conundrum)

Cheryl Teich
Teich Process Development LLC
AIChE Process Development Symposium
Philadelphia, PA
8 June 2022

ACCELERATED MODULAR PROCESS DEVELOPMENT (AMPD) FOR DOMESTIC MANUFACTURING OF CRITICAL ACTIVE PHARMACEUTICAL INGREDIENT CHEMICAL PRECURSORS

NIST Rapid Assistance (for) Coronavirus Economic Response (**RACER**) Grant

RAPID® Transforming Process Industries

Before we get started, some audience participation

Continuing with the Audience Participation

Who has heard of RAPID?

Who is willing to share their understanding of RAPID?

Homework (that means later):

Check out RAPID's 2 minute, 13 second
YouTube video on Process Intensification

RAPID RACER Project Outcomes

1. Development and dissemination of an accelerated process development framework that takes advantage of standardized modular process operations.

2. Setup and operation of a modular process development Center of Excellence (COE) at MATRIC in South Charleston, WV, that will serve as a national asset testbed for future coronavirus response and a hub for job creation in the underserved Appalachian region.

3. Demonstration of the accelerated process development framework for a coronavirus relevant API precursor specialty chemical.

   Designation used internally at MATRIC: SAPPHIRE (Supplying Appalachian Produced PHarmaceutical Intermediates from Research Excellence)

Why SAPPHIRE?

Sapphires are known as the stones of wisdom and serenity. They are used to release mental tension and dispel depression. They are believed to facilitate meditation, calm the mind, and **provide the mental clarity required to surmount obstacles**.

Presenting the Facets (oops, Faces) of SAPPHIRE

Teich Process Development LLC

Jack Dever

MATRIC

Cheryl Teich

PROCEGENCE

Ignasi Palou-Rivera

Mike Burgess

Rob Nunley

Nima Yazdanpanah

Project Impact, Outcomes and Dissemination

The project will immediately impact coronavirus response through creation and testing of a new accelerated methodology and modular processing COE for development of processes to manufacture API specialty chemical precursors. ***Tools and capabilities such as this do not currently exist in the U.S. for general use. The process development framework and national asset testbed capabilities will be made available to RAPID's members and disseminated broadly to AIChE's Process Development Division and Pharmaceutical Discovery, Development and Manufacturing Forum (PD2M).*** The team envisions that the COE will also serve as a tool for SMM to develop processes and build production capacity for on-shoring of selected specialty chemicals, API precursors, and APIs, and the COE/MATRIC may serve as a production center for future coronavirus response scenarios.

Project Tasks

R:RAPID M: MATRIC T: TEICH PROC. DEV. P: PROCEGENCE

Task	Description	Timing	Lead
1.1	Gather and codify historic information related to chemical process development stages and critical deliverables.	M1-3	T
1.2	Assess historic data and perform gap analysis to determine critical stages for acceleration of process development.	M4-6	T
2.1	Define stagewise methodology for accelerated process development.	M6-9	T/M/R
2.2	Define equipment infrastructure and modular testbed needs.	M6-12	R
3.1	Select precursor chemistries most relevant to coronavirus response.	M1-3	P
3.2	Select specialty chemical precursor candidate(s) and perform initial process development assessment.	M4-9	P
4.1	Modular Process Development Equipment Identification and Design	M1-3	M/T/R
4.2	Prioritize and Specify Laboratory and Modular Pilot Plant Systems.	M4-8	M
4.3	Procure Equipment and Construct/Commission Pilot Plant Modules.	M4-15	M
5.1	Complete Laboratory Studies for Baseline Process Understanding	M7-15	M
5.2	Develop Conceptual Design for Proposed Process.	M16-18	M
5.3	Complete Pilot Plant Process Validation.	M19-24	M
5.4	Propose Commercial Scale Process Design.	M22-24	R
6.1	Report project progress on routine basis.	M1-24	R
6.2	Disseminate project results to process development community.	M1-24	R

RAPID®
Transforming Process Industries

Consider the yellow brick road from idea to profit

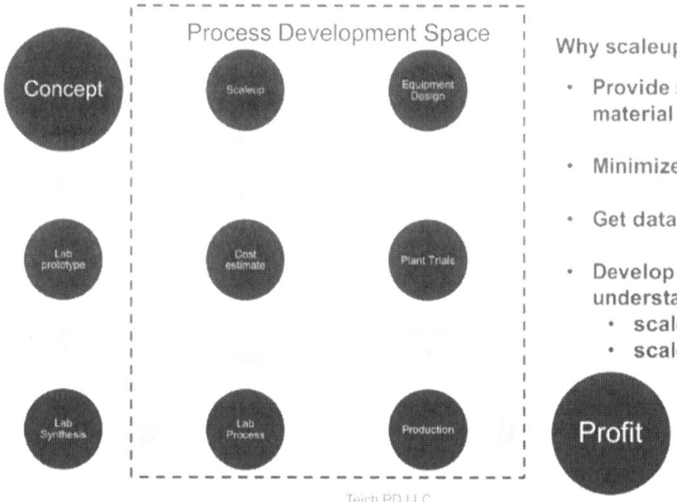

Why scaleup?

- Provide sample and test material
- Minimize risk
- Get data
- Develop process understanding
  - scale-independent factors
  - scale-dependent factors

RAPID®
Transforming Process Industries

SAPPHIRE'S Goal is to Decrease Time to Profit without Sacrificing Understanding

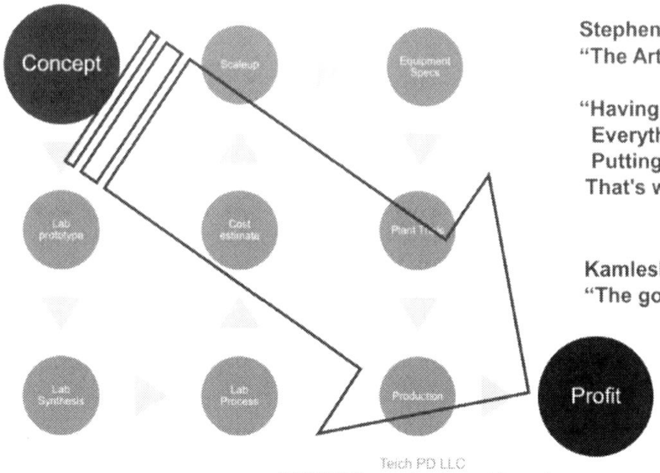

Stephen Sondheim
"The Art of Making Art"

"Having just a vision's no solution,
Everything depends on execution:
Putting it together-
That's what counts!"

Kamlesh Bhatia, DuPont (ret.)
"The goal is commercial success."

RAPID®
Transforming Process Industries

Develop a Process Understanding Practice

Follow **Don Koestler*'s** First, Second, and Third Laws

Warrior 1 Warrior 2 Warrior 3

Know Your Process Know Your Process Know Your Process

Obey **Makac's** rule: "Don't assume nothing!"

Scale-up Systems states it more elegantly: "Question Everything!"

**My dearly departed Rohm and Haas mentor. May his memory be a blessing.*

RAPID®
Transforming Process Industries

Develop a Process Understanding Practice

Alternatively, heed Mike Schultz's advice. The idea must

Warrior 1	Warrior 2	Warrior 3
Meet commercial objectives	Obey laws of thermodynamics and economics	Attain Mass and energy balance; Product Value > RM Cost (Douglas Level 1)

Source: Mike Schultz, "Develop the Right Innovative Concepts Using Early Stage Process Design, AIChE 2012 Process Development Symposium

RAPID® Transforming Process Industries

Framework for RACER Test Bed Process Development and Design

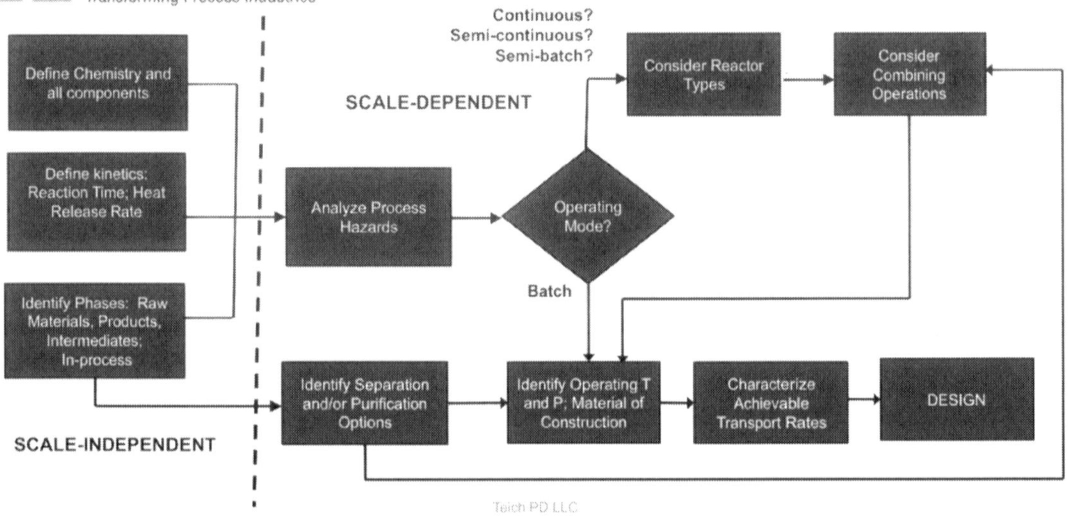

Teich PD LLC
AIChE 2022 Process Development Symposium
8 June 2022
13

RAPID® Transforming Process Industries

Accelerated PD: Overlay Reaction Process Understanding with Modular KB

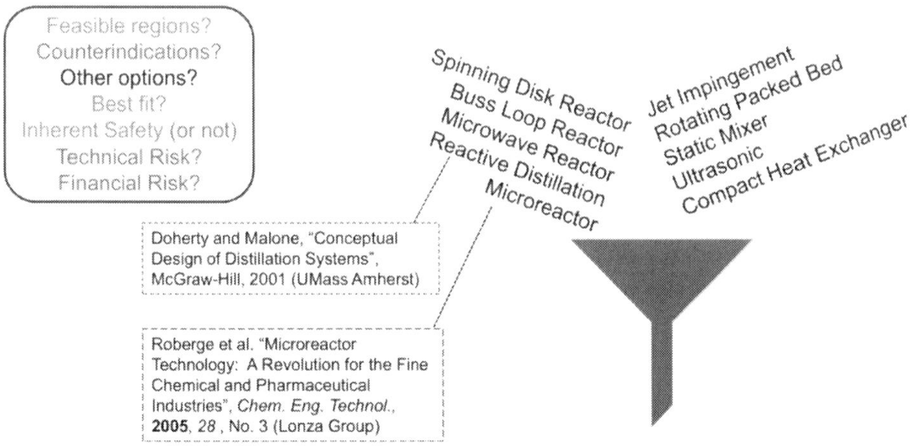

Teich PD LLC
AIChE 2022 Process Development Symposium
8 June 2022
14

Accelerated PD: Overlay Separation Process Understanding with Modular KB

RAPID® Transforming Process Industries

Feasible regions?
Counterindications?
Other options?
Best fit?
Inherent Safety (or not)
Technical Risk?
Financial Risk?

Lawton, "Rapid Scale-up of pharmaceutical processes: Finding the trees in the forest", EECE-6 Proc., Copenhagen, 16 – 20 Sept 2007

Distillation
Liquid Extraction
Precipitation
Crystallization

Atherton and Carpenter, "Process Development: Physicochemical Concepts", Oxford University Press, 2005 reprint

My Favorite Quotes of the First PDS

RAPID® Transforming Process Industries

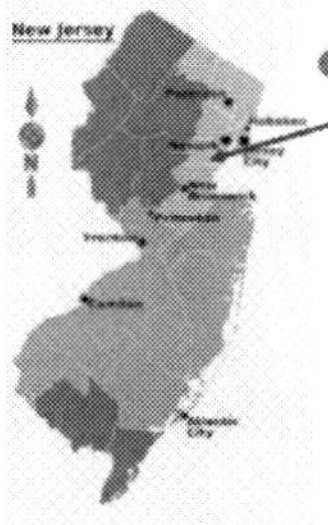

MERCK

- Dr. Arthur Andrews, VP of Process R&D, Merck
  - Don't expect
    - Robust chemistry to overcome poor process technology
    - Innovative engineering to compensate for poor chemistry!
  - Beware of chemists bearing Waring blenders!
- Ed Paul (Z"L), Chemical Engineering R&D Department, Merck (ret.)
  - Always assume there is a mixing problem until proven otherwise.
  - Laboratory crystallization can GOOD, BAD, or UGLY.
  - It May Not Be Possible to Do Every Operation in a Tank

Where are you on the Xerox PD Model?

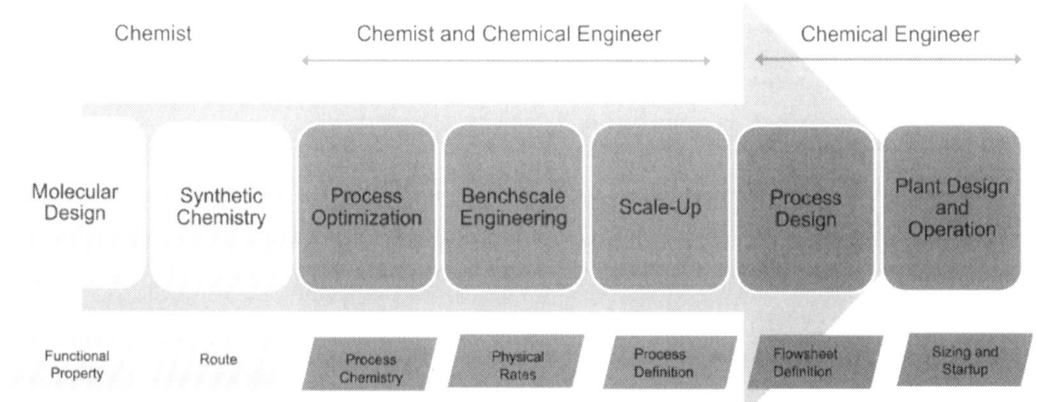

Source: Paul Szabo, Xerox, "The Best Technology Transfer Steps Are: NO TECHNOLOGY TRANSFER STEPS", AIChE Process Development Symposium, June 2003

Where are you in the RandH Ongoing Process Development Work Process?

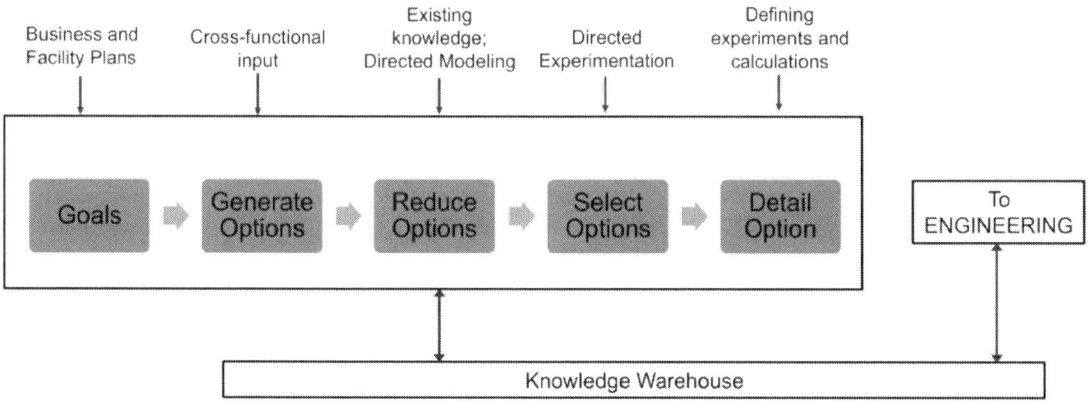

Source: Aaron Sarafinas, Rohm and Haas Company, "Successful Process Development: Profitable Specialty Chemical Products Using Winning Work Processes, AIChE Process Development Symposium, June 2003

RAPID® Transforming Process Industries

Please Take our SHORT Survey

1. Define your Process Development Space
2. Assign Roles/Responsibilities
3. Provide your contact information if you are interested in participating further in SAPPHIRE Task 1

RAPID® Transforming Process Industries

Draw Your Process Development Space

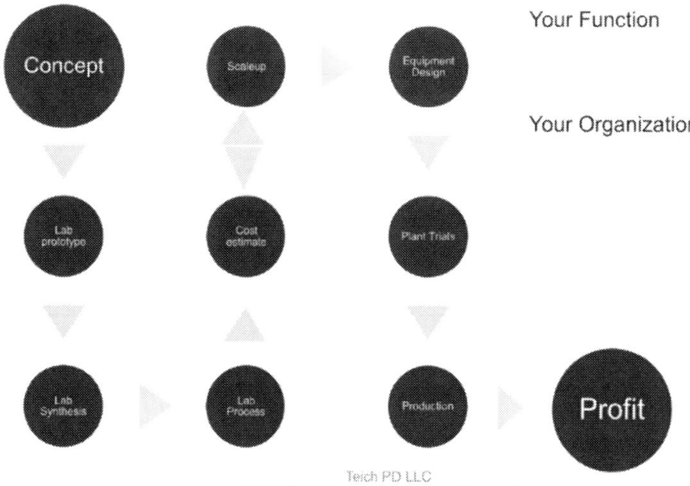

Your Function

Your Organization type

RAPID® Transforming Process Industries

Assign Roles/Responsibilities

Your Function/Role

Your Organization type

RAPID® Transforming Process Industries

Provide Contact Information (Optional) to Provide Additional SAPPHIRE Task 1 Input

- Name:
- Organization
- Role/Job title
- Time in current role
- Time in current organization
- Email
- Phone

Scan to link to on-line Contact Info form

RAPID®
Transforming Process Industries

THANK YOU!

> AIChE Process Development Symposium
> Mune 2022
> Darren Youngs
> Process Development Manager

Recycling Carbon Through Gas Fermentation

LanzaTech

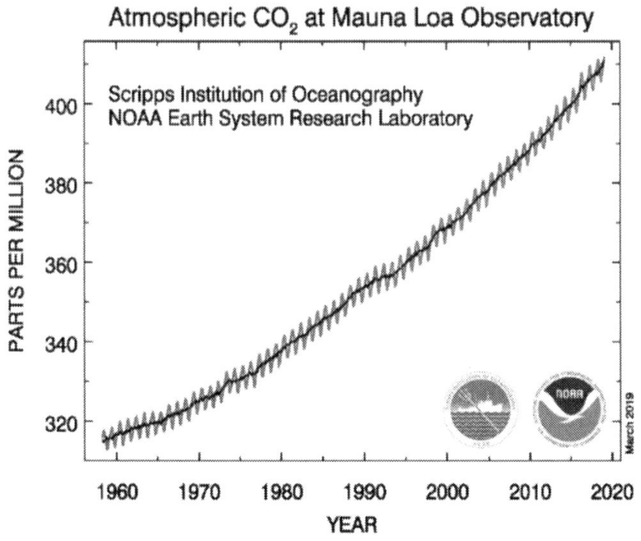

"It is unequivocal that human influence has warmed the atmosphere, ocean and land. Widespread and rapid changes in the atmosphere, ocean, cryosphere, biosphere have occurred" - IPCC's 6th Assessment Report on Climate Change

LanzaTech 2

It’s time to rethink carbon

Every waste is a resource

LanzaTech 5

Including CO_2

LanzaTech

Predictions are simply extrapolations of the past...

...innovation expands the 'art of the possible'

LanzaTech 7

Gas Fermentation

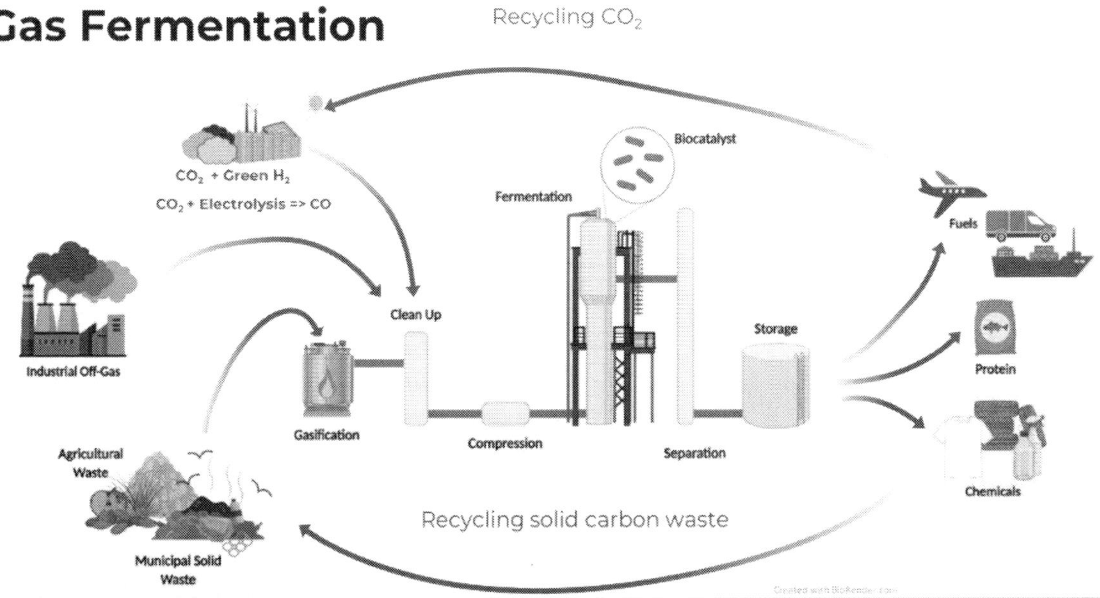

LanzaTech

Harnessing Biology

95% selectivity to ethanol across different gas compositions

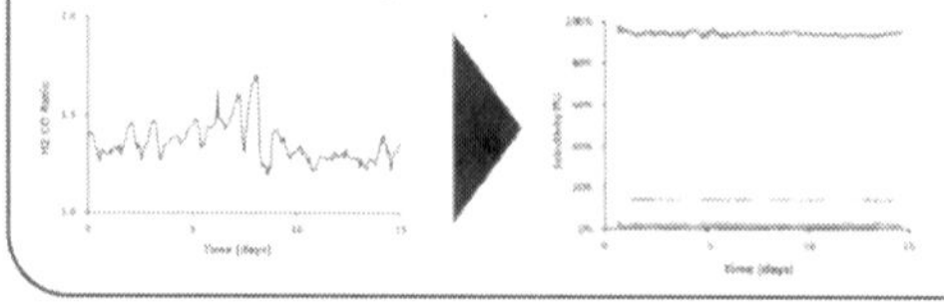

- Self replicating
- Self repairing

Incorporates a water gas shift reaction

$CO + H_2O \leftrightarrow CO_2 + H_2$

$[H_2]$

- Operates at 37°C
- Contaminant tolerant
- Consumes sulfur

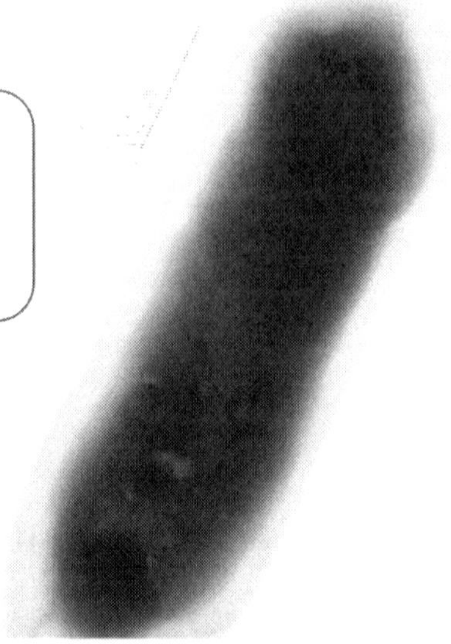

LanzaTech 9

Understanding The Potential

	Reaction	H_2:CO Ratio	Carbon Efficiency
CO	$6\ CO + 3\ H_2O \rightarrow$ <u>EtOH</u> $+ 4\ CO_2$	0:1	33.3%
CO + H_2	$3\ H_2 + 3\ CO \rightarrow$ <u>EtOH</u> $+ CO_2$	1:1	66.7%
CO + H_2	$4\ H_2 + 2\ CO \rightarrow$ <u>EtOH</u> $+ H_2O$	2:1	100%
CO + H_2 + CO_2	$5\ H_2 + 1\ CO + 1\ CO_2 \rightarrow$ <u>EtOH</u> $+ 2\ H_2O$	5:1	100%

Flexible conversion platform
Multiple routes to 100% carbon conversion

LanzaTech

Quantifying The Potential

Ethanol Potential of Gas Fermentation Feedstocks

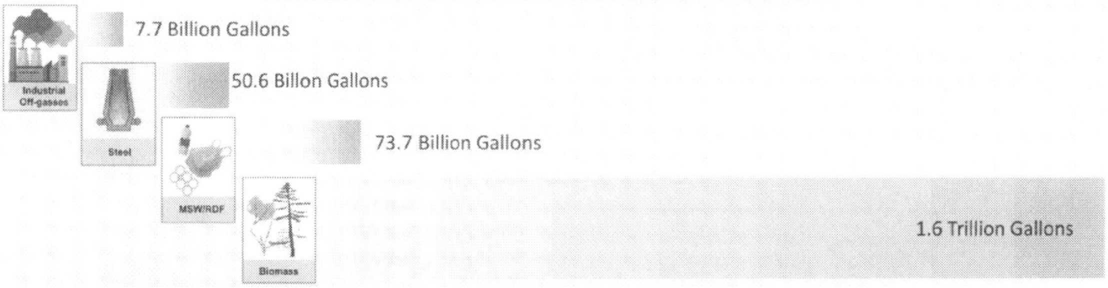

- Waste gasses represent an enormous opportunity for carbon capture and transformation
- Over 1.6 trillion gallons of potential ethanol is currently stored in solid wastes such as MSW or agricultural byproducts
- Practically limitless quantify of carbon (~2.1 gigatons of carbon) is currently present in the atmosphere

Accessing The Potential

- Characterizing potential feedstocks is critical to derisking commercial deployments
- Each gas source represents unique challenges for downstream processing
- LanzaTech identified the need to develop a portable laboratory to analyze gas sources in real-time
  - Gas Testing Station (GTS) – Integrated analytics and gas fermentation
  - Gas Analysis System (GAS) – Dedicated analytical system

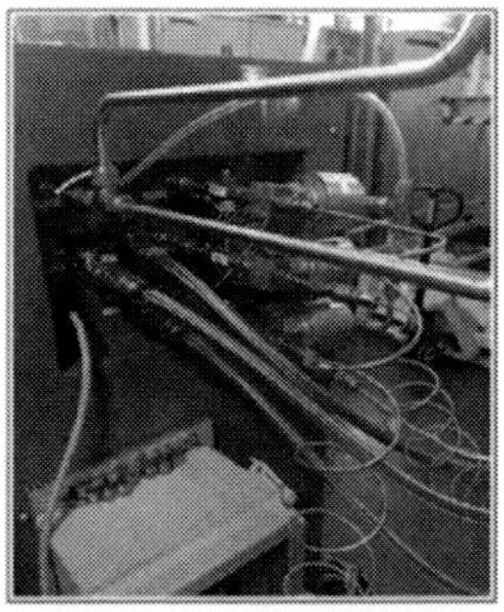

Competitive Advantage

Biological processes have inherent advantages over thermo-catalytic processes

- ***Thermo-catalytic processes*** – catalyst poisoning compromise performance until catalyst is regenerated or replaced.
- ***Biological processes*** – Self-regenerating once inhibitor is removed

Fischer-Tropsch Catalyst – S Poisoning

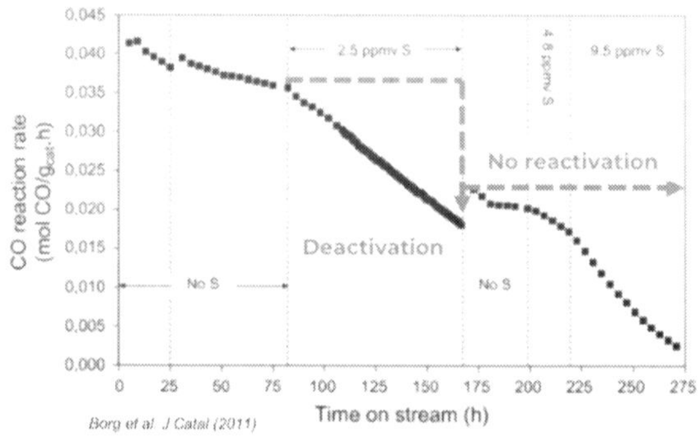

Borg et al. J Catal (2011)

Biocatalyst – O_2 Inhibition

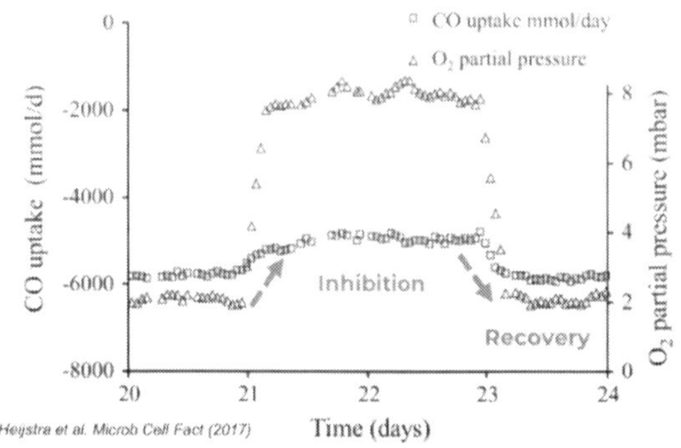

Heijstra et al. Microb Cell Fact (2017)

LanzaTech 13

Path to Commercialization

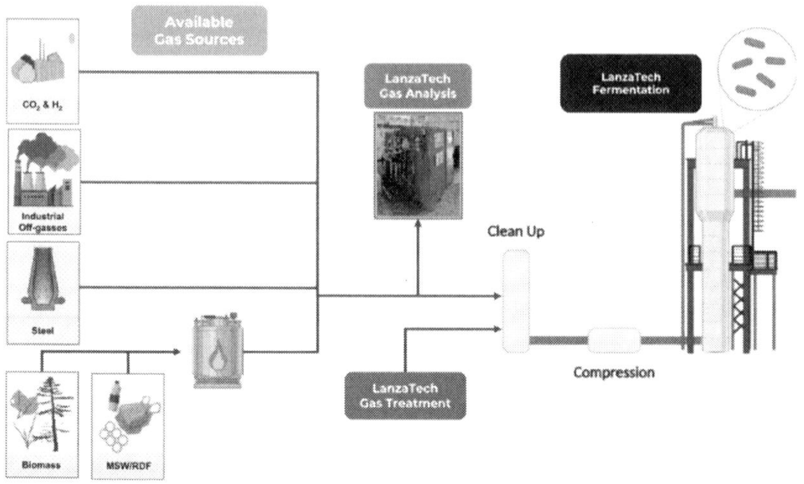

LanzaTech

LanzaTech

What was said:

1992	All gas fermenting acetogens make acetate as their sole product
2000	Gas fermentation of CO/H_2 is inherently mass transfer limited preventing commercial use

15

Commercial-Scale Continuous Operation

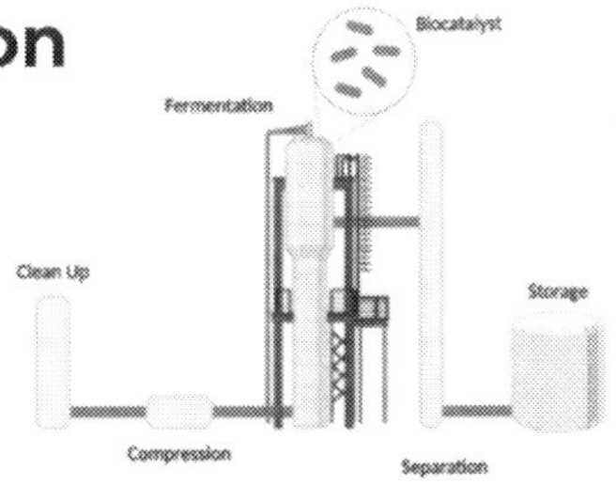

Stable, continuous performance

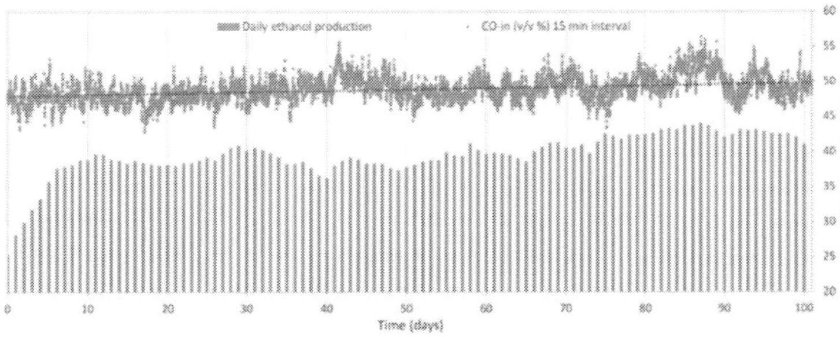

LanzaTech

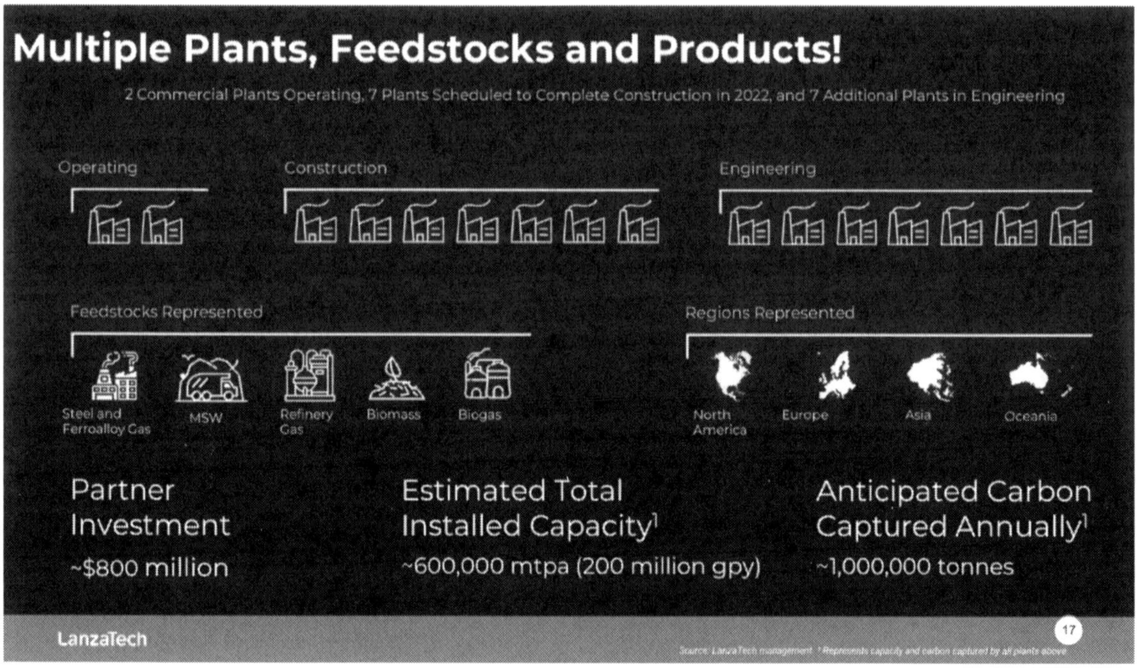

What was said:

1992	All gas fermenting acetogens make acetate as their sole product
2000	Gas fermentation of CO/H_2 is inherently mass transfer limited preventing commercial use
2009	Gas fermenting acetogens are genetically inaccessible
2016	The complexity of working with anaerobic acetogens is too high for high-throughput engineering to become possible
2019	Other than acetate or ethanol it is impossible to make any other product at high selectivity through anaerobic gas fermentation

LanzaTech

18

Fundamental Research

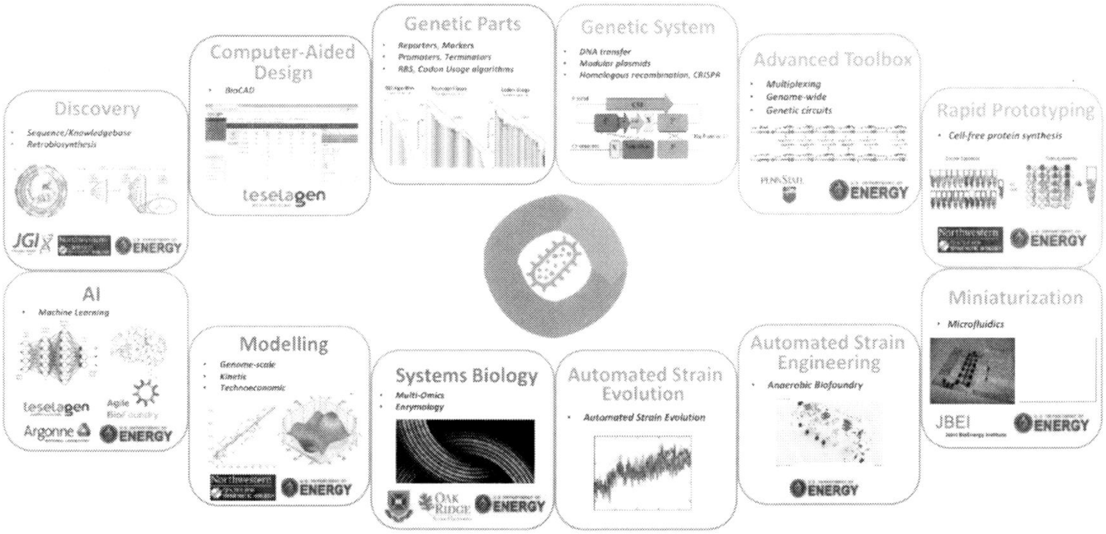

LanzaTech

Advancing Science

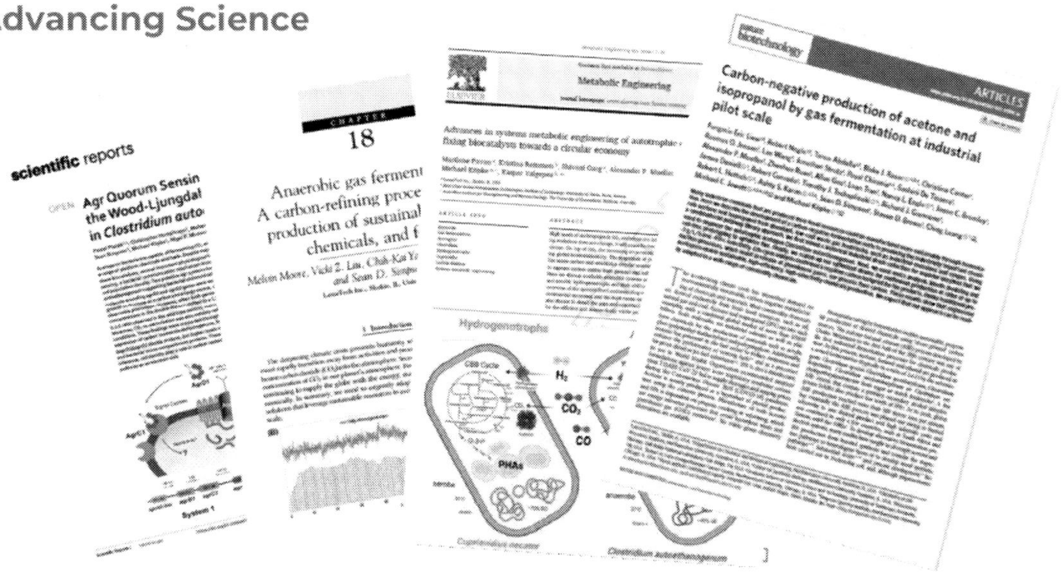

LanzaTech 20

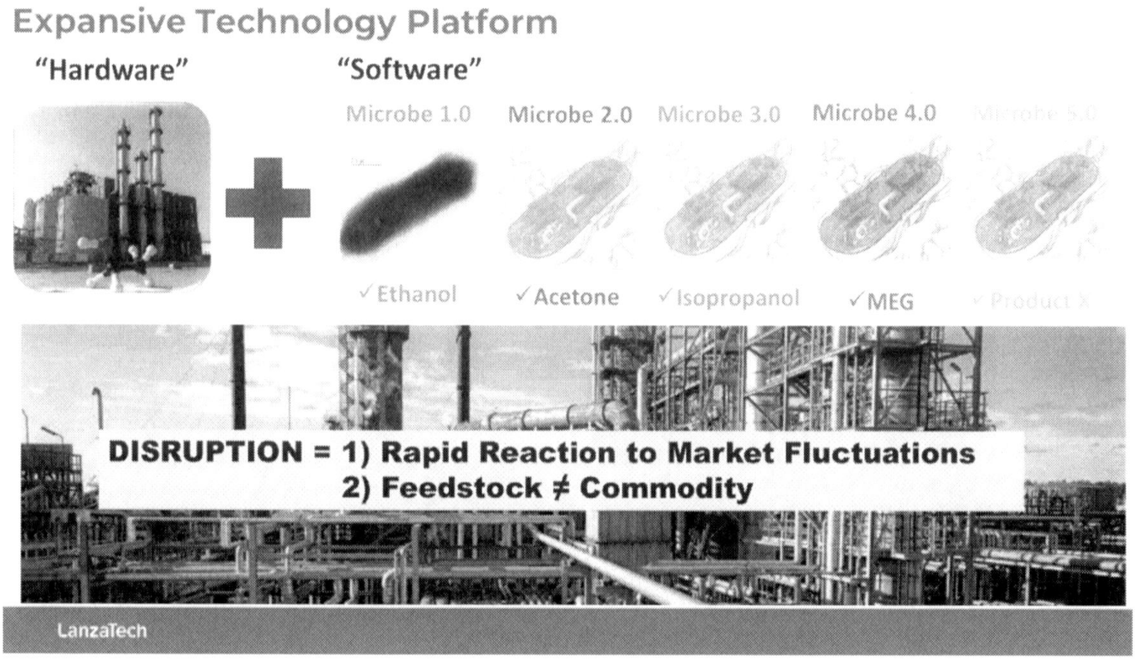
Expansive Technology Platform
"Hardware"
"Software"
Microbe 1.0
Microbe 2.0
Microbe 3.0
Microbe 4.0
✓Ethanol
✓Acetone
✓Isopropanol
✓MEG
DISRUPTION = 1) Rapid Reaction to Market Fluctuations
2) Feedstock ≠ Commodity
LanzaTech

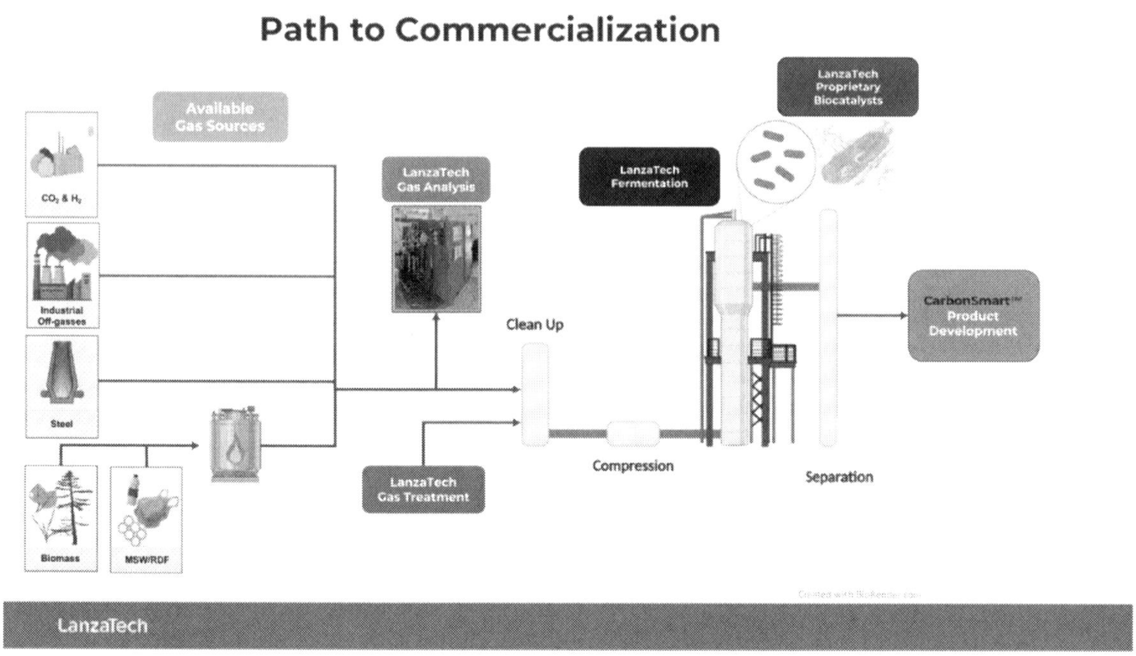
Path to Commercialization
Available Gas Sources
CO₂ & H₂
Industrial Off-gasses
Steel
Biomass
MSW/RDF
LanzaTech Gas Analysis
LanzaTech Gas Treatment
Clean Up
Compression
LanzaTech Fermentation
LanzaTech Proprietary Biocatalysts
Separation
CarbonSmart™ Product Development
LanzaTech

lululemon Unilever L'OREAL mibelle GROUP MIGROS ZARA dm

LanzaTech

Carbon Emissions to Sustainable Aviation Fuels

Abundant, Waste-based Feedstock + Low-Cost Process = Competitive Waste-to-SAF Solution

With opportunity to utilize existing ethanol supply today

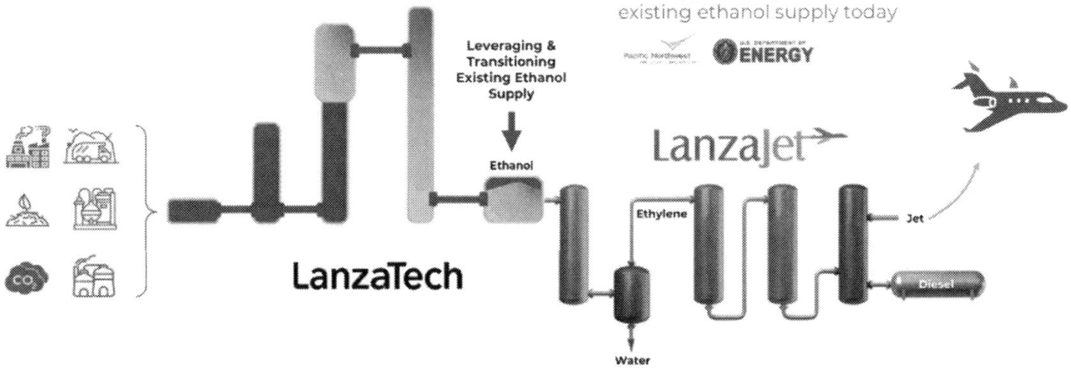

Source: LanzaTech management

LanzaTech

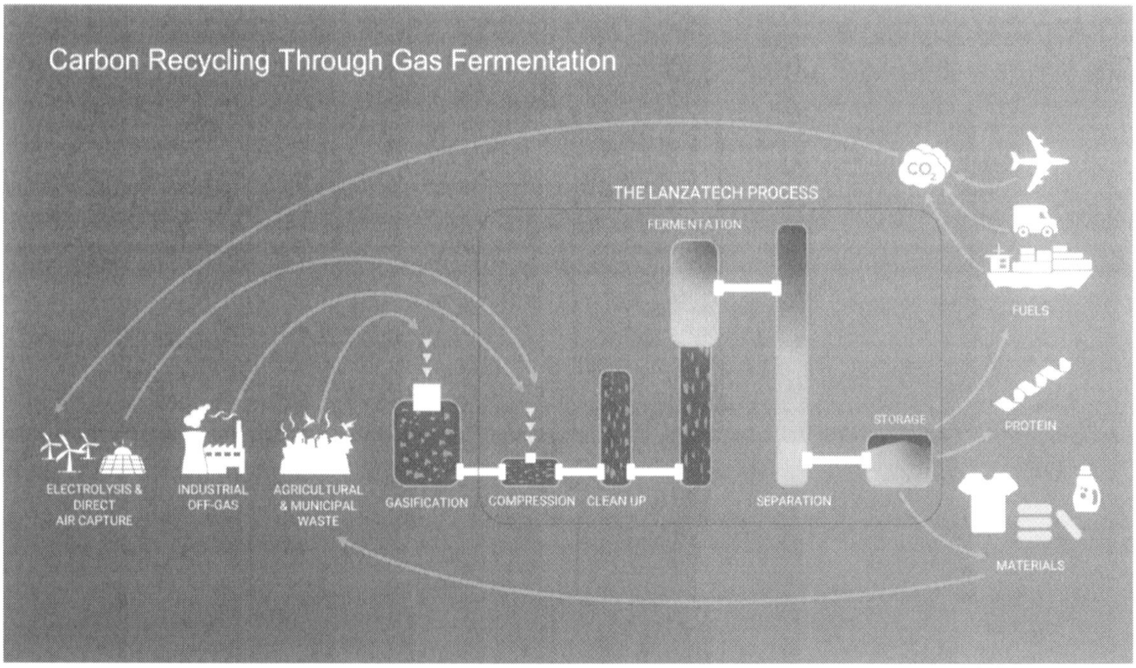
Carbon Recycling Through Gas Fermentation
THE LANZATECH PROCESS
FERMENTATION
CO_2
FUELS
PROTEIN
STORAGE
MATERIALS
ELECTROLYSIS & DIRECT AIR CAPTURE
INDUSTRIAL OFF-GAS
AGRICULTURAL & MUNICIPAL WASTE
GASIFICATION
COMPRESSION
CLEAN UP
SEPARATION

Thank you!
LanzaTech

CORTEVA agriscience

Sustainability Metrics for Small Molecule Development

Abe Schuitman, Process Development Engineering Team Leader

June 2022

Sustainability in Small Molecule Development

- Corteva 2030 Sustainability Goals
- Tools used in Process Development
- Sustainability Scorecard – Why is this important?
  - "If you can't measure it, you can't fix it" (i.e. know if we're winning or losing)
  - Data driven decision making – compare/contrast/select
- Route Selection Criteria

- Sustainability needs to be built into active ingredients & critically evaluated

BUILDING A SUSTAINABLE FUTURE

https://youtu.be/m2Z8lFp2rCQ

CORTEVA agriscience

Approved for External Audiences

R&D Sustainable Innovation

- Every newly developed Corteva Agriscience innovation from our pipeline will meet our sustainability criteria by 2025
- Sustainability assessed and tracked at every stage and applies to all pipeline projects.

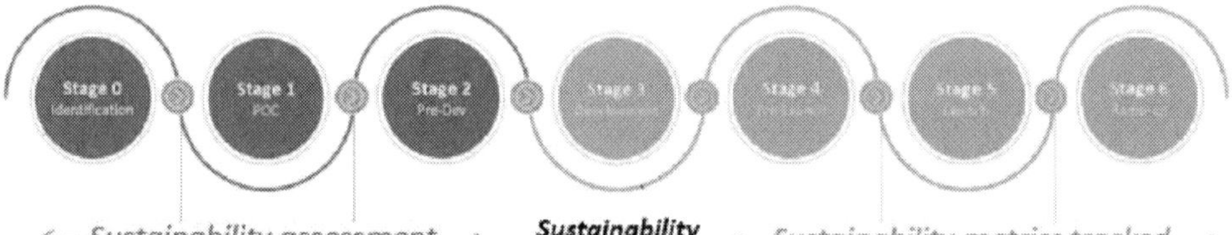

CORTEVA agriscience

4

Sustainable Innovation Criteria

* Meets baseline requirements, delivers at least one notable sustainability advantage, while maintaining parity comparted to other products in the market today

CORTEVA agriscience

5

Corteva Tools/Process During Discovery & Development Phases

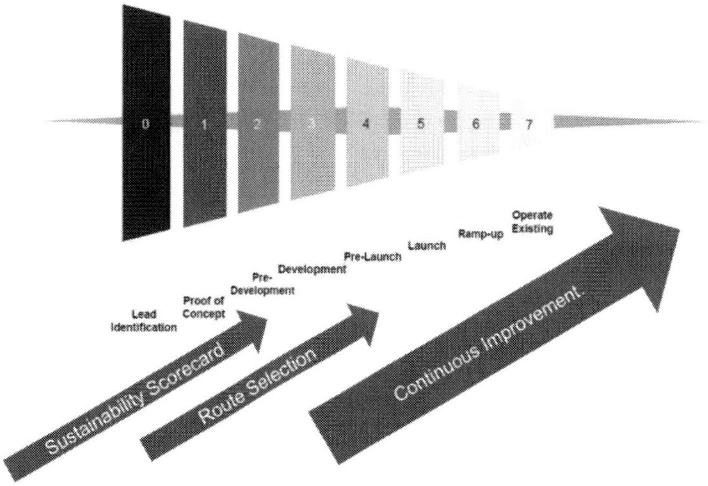

Leading Indicators

- ✓ Guide Research
- ✓ Make Decisions
- ✓ Drive Holistic Process Improvements

CORTEVA agriscience

6

Scorecard:
Sustainable Active Ingredient Manufacture

Seven categories prioritize and guide research:

1. Process Mass Intensity:
   minimize waste and maximize atom efficiency
2. Solvents:
   use benign "green" solvents
3. Toxicity:
   avoid highly toxic chemicals
4. Energy Release Potential:
   avoid highly energetic chemistries
5. Temperature Range:
   minimize energy input during processing
6. Metals:
   minimize use of heavy metals & transition metals
7. Renewable Content:
   use renewable (non-petroleum) feedstocks

Metrics align to American Chemical Society Green Chemistry principles

Sustainability Status

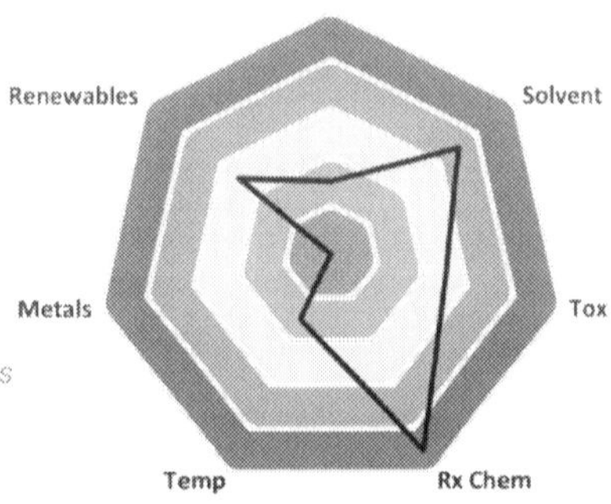

The 12 Principles of Green Chemistry

- ✓ Prevent Waste
- Design Safer Chemicals and Products
- ✓ Design Less Hazardous Chemical Syntheses
- ✓ Use Safer Solvents/ Reaction Conditions
- ✓ Increase Energy Efficiency
- ✓ Use Renewable Feedstocks
- Design Chemicals and Products that Degrade After Use
- ✓ Minimize the Potential for Accidents
- Analyze in Real Time to Prevent Pollution
- ✓ Use Catalysts, Not Stoichiometric Reagents
- ✓ Maximize Atom Economy
- ✓ Avoid Chemical Derivatives

CORTEVA agriscience 7

Toxicity: Identifying a metric that spans R&D + Manufacturing

- R&D safety criteria has been standardized on the Globally Harmonized System (GHS) i.e. H310/H330
  - Hazards & Review criteria input available on each SDS
- Manufacturing utilizes very different metrics centered around LOPA/RAST analyses
- Multiple proposals to push towards the following:
  1. Differentiate between toxicity of reagents used in an **industrially relevant manner**
     - Toxicity combined with Risk/Exposure potential
  2. Differentiate between reagents with regard to their complexity to handle at commercial scale.
     - Use reliable data sources that are publicly available
     - Low activation barrier for usage

Green Chemistry Principles #3 (Less Hazardous Syntheses) and #12 (Inherently Safer Chemistry)

CORTEVA agriscience 8

Toxicity Scales/Indexes Evaluated

1. Hodge and Sterner Toxicity Scale*
   - Key inputs: LD_{50}/LC_{50}
   - Correlates LD_{50} per kg body weight with Degree of Toxicity
2. Research Toxicity Index (RTI)
   - Key inputs: Vapor Pressure, Molecular Weight, ERPG-2

$$RTI = \sqrt{\frac{20000(VP)}{MW(ERPG\text{-}2)}}$$

3. Chemical Exposure Index (CEI)**
   - Key inputs: ERPG-2, Airborne Quantity, Distance of Concern
4. Substance Hazards Index (SHI)***
   - Key inputs: PAC-3, Vapor Pressure

$$SHI = \frac{Vapor\ Pressure\ @\ 20-25\ ^{\circ}C(mmHg)}{760} * \frac{10^{\wedge}6}{PAC\ 3\ (ppm)}$$

*N. Irving Sax, Dangerous Properties of Industrial Materials (New York: Van Nostrand Reinhold, 1984), p. 1.

**https://www.aiche.org/ccps/resources/glossary/process-safety-glossary/dow-chemical-exposure-index-cei

***API RP 750, Management of Process Hazards, First Edition,

CORTEVA agriscience

Toxicity Scales/Indexes Evaluated

1. Hodge and Sterner Toxicity Scale*
   - Key inputs: LD_{50}/LC_{50}
   - Correlates LD_{50} per kg body weight with Degree of Toxicity

- ☑ Differentiates toxicity
- ⃠ No risk/exposure variable
- ⃠ LC_{50} data not easily accessible (units!)

2. Research Toxicity Index (RTI)

Routes of Administration			
Toxicity Rating	Commonly Used Term	Inhalation LC_{50} (exposure of rats for 4 hours) ppm	Probable Lethal Dose for Human
1	Extremely Toxic	10 or less	1 grain (a taste, a drop)
2	Highly Toxic	10-100	4 ml (1 tsp)
3	Moderately Toxic	100-1000	30 ml (1 fl. oz.)
4	Slightly Toxic	1000-10,000	600 ml (1 pint)
5	Practically Non-toxic	10,000-100,000	1 litre (or 1 quart)
6	Relatively Harmless	100,000	1 litre (or 1 quart)

*N. Irving Sax, Dangerous Properties of Industrial Materials (New York: Van Nostrand Reinhold, 1984), p. 1.

**https://www.aiche.org/ccps/resources/glossary/process-safety-glossary/dow-chemical-exposure-index-cei

***API RP 750, Management of Process Hazards, First Edition,

CORTEVA agriscience

Toxicity Scales/Indexes Evaluated

2. Research Toxicity Index (RTI)

$$RTI = \sqrt{\frac{20000(VP)}{MW(ERPG\text{-}2)}}$$

- Key inputs: Vapor Pressure, Molecular Weight, ERPG-2

- ☑ Differentiates toxicity (ERPG-2)
- ☑ Includes risk/exposure variable (i.e. VP)
- ⃠ ERPG-2 values are available, but limited

*N. Irving Sax, Dangerous Properties of Industrial Materials (New York: Van Nostrand Reinhold, 1984), p. 1.

**https://www.aiche.org/ccps/resources/glossary/process-safety-glossary/dow-chemical-exposure-index-cei

***API RP 750, Management of Process Hazards, First Edition,

Toxicity Scales/Indexes Evaluated

3. Chemical Exposure Index (CEI)**

- Key inputs: ERPG-2, Airborne Quantity, Distance of Concern

- ☑ Differentiates toxicity (ERPG-2)
- ☑ Includes risk/exposure variable (i.e. airborne quantity)
- ⃠ ERPG-2 values are available, but limited
- ⃠ Additional unknowns unapplicable for R&D (i.e. airborne & distance variable)

*N. Irving Sax, Dangerous Properties of Industrial Materials (New York: Van Nostrand Reinhold, 1964), p. 1.

**https://www.aiche.org/ccps/resources/glossary/process-safety-glossary/dow-chemical-exposure-index-cei

***API RP 750, Management of Process Hazards, First Edition,

Substance Hazards Index (SHI) Meets Criteria

$$SHI = \frac{Vapor\ Pressure\ @\ 20-25\,°C(mmHg)}{760} * \frac{10^{\wedge}6}{PAC\ 3\ (ppm)}$$

1. Elements of both toxicity & risk
2. Protective Action Criteria (PAC) tabulated & broadly available on-line through the US DOE
   - https://edms3.energy.gov/pac/Search
3. Differentiated between highly hazardous reagents
   - Phosgene (H330) SHI = ~2,500,000 versus Acetic Anhydride (H330) SHI = 155

Sustainability Scorecard Example (Stage 0-1): Higher Score is Better

**Industrial standard

**Relevancy in R&D and manufacturing

Route A

Sustainability Category	Score
Process Mass Intensity	**15**
Solvent Selection	9
Toxicity of worst substance	15
Reactive Chem	-10
Temperature range	7
Precious metals	**16**
Total Score	**43**

- Higher PMI
- Eliminated Precious Metals

Route B

Sustainability Category	Score
Process Mass Intensity	**30**
Solvent Selection	9
Toxicity of worst substance	15
Reactive Chem	-10
Temperature range	7
Precious metals	**-10**
Total Score	**41**

- Lower PMI
- Usage of Precious Metals

Route Selection Criteria

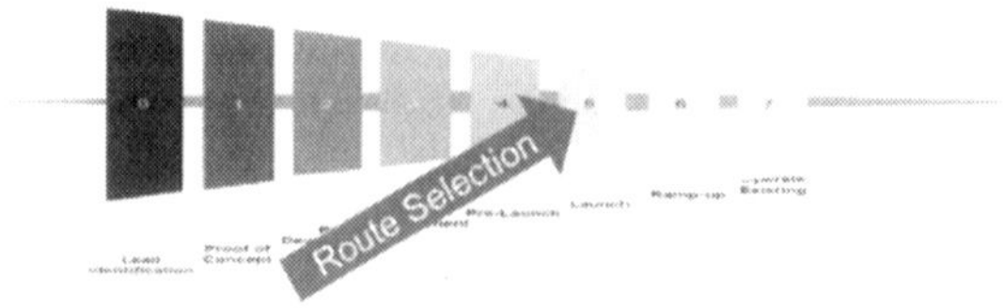

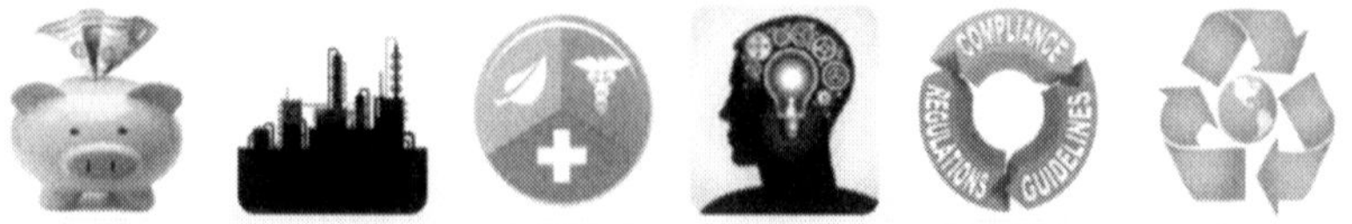

COM	Manufacturing Complexity	EH&S	IP	Regulatory & Biology	Sustainability
Overall Yield Raw Material Cost Raw Material Availability	Process Complexity Chemistry Difficulty Process Robustness Type of Unit Operations	Waste Treatment Industrial Hygiene Reactive Chemicals Air Emissions	Freedom to Operate	Quality and Specifications	PMI Solvents Toxicity Energy release Temperatures Precious metals Renewability

CORTEVA agriscience 15

Route Selection Criteria and Correlation

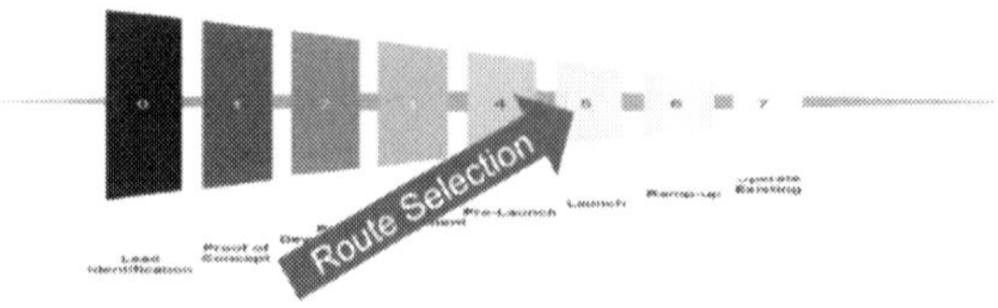

COM
Overall Yield
Raw Material Cost
Raw Material Availability

- Can the project team achieve the business objective with the routes under consideration?
- COM (Cost of Manufacturing) trends with Sustainability indicators:

↓ **PMI Reductions** ~ ↓ **Lower COM**

↓ **Safer Rx Chem** ~ ↓ **Lower COM**

↓ **Improved Tox** ~ ↓ **Lower COM**

↓ **Eliminate Precious Metals** ~ ↓ **Lower COM**

CORTEVA agriscience 16

Route Selection Criteria – Manufacturing Complexity

Manufacturing Complexity
Process Complexity
Chemistry Difficulty
Process Robustness
Type of Unit Operations

- Challenging questions centered on the ability for the selected chemistry to translate into process
- Specialty equipment requirements that could limit where production can take place
- High level 'fit' into envisioned manufacturing assets

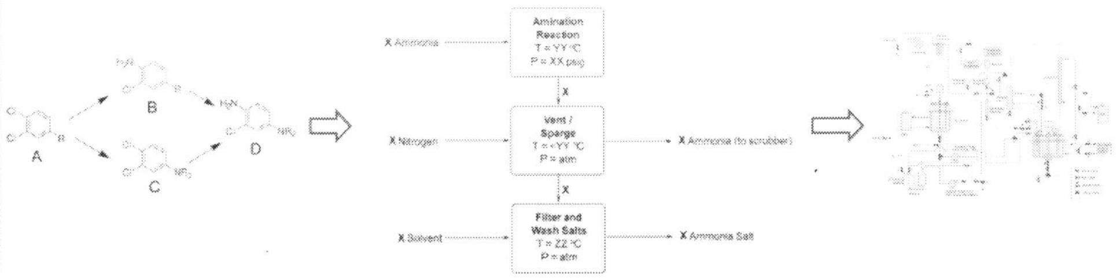

Route Selection Criteria – EH&S

EH&S
Waste Treatment
Industrial Hygiene
Reactive Chemicals
Air Emissions

- All waste is not created equally (go beyond a singular PMI #)
  - More details process information allows for a more thorough comparison (organic/aqueous/solid)
  - "Show-stopping" issues regarding waste are identified
  - Detailed discussion conducted with our manufacturing teams and Subject Matter Experts (SME)
- Reactive chem analysis complete with key hazards identified
  - Again, a key focus on "show-stopping" issues has typically been completed
  - Areas of significant engineering controls also identified

Route Selection Criteria – Sustainability

Green Chemistry #6 – Increase Energy Efficiency

Green Chemistry #12 – Minimize Accidents

Cryogenic conditions can significantly add to energy consumption

Sustainability
PMI
Solvents
Toxicity
Energy release
Temperatures
Precious metals
Renewability

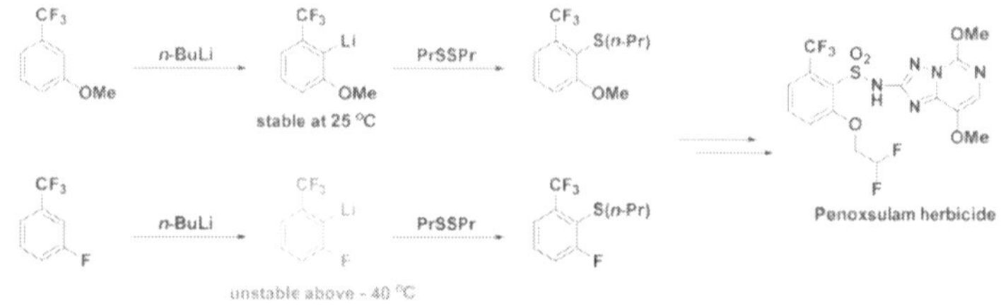

Benzyne formation through Li-F elimination

Leng, *et al. Org. Proc. Res. Dev.* **2012**, *16*, 415

CORTEVA agriscience 19

Route Selection Criteria – Holistic Evaluation

Green Chemistry #1 – Prevent Waste

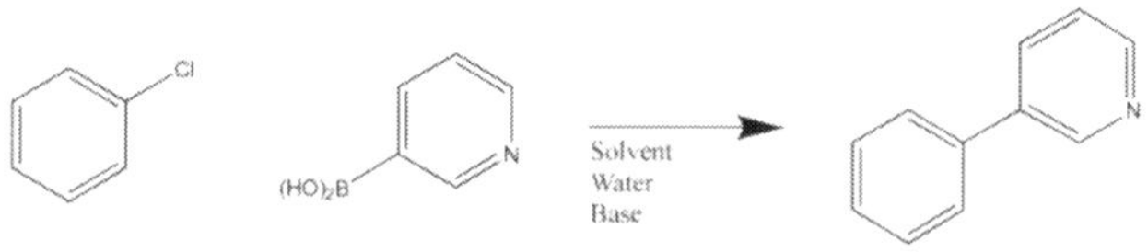

EH&S
Waste Treatment
Industrial Hygiene
Reactive Chemicals
Air Emissions

Base Selection Option 1:
- Inorganic phosphate base

↑ Yield ↑ Reactive Chem

Showstopper: Phosphate waste treatment

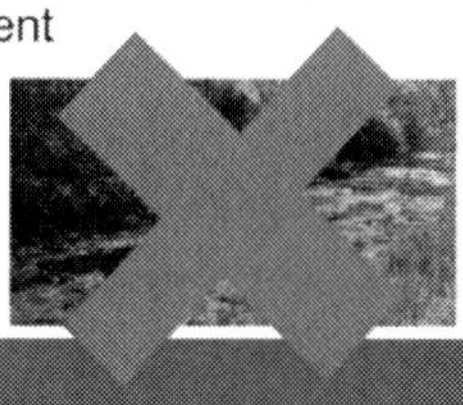

Base Selection Option #2: ☑
- Organic amine base

↓ PMI ↓ Cost

- Recycle strategy implemented around azeotropes
- Enabled water recycle in addition to the solvent & base
- Recovered >280 mtons of base/solvent /water during Campaign #1

CORTEVA agriscience 20

Route Selection (Example Stage 3)

	Route 1	Route 2	Route 3
Conversion Step Count	7	6	8
Unit Operations	31	30	37
Hydraulic Throughput *	0.18 kg / gal	0.18 kg / gal	0.18 kg / gal
Cost of Manufacture	$200-260	$180-230	$220-270
Process Robustness	Medium	High	Low
Unique EHS Concern	No	Acetyl Chloride	DMAP
Process Understanding	High	High	Low
PMI	74	61	93
Overall Yield	28%	58%	43%

☑ (Route 2)

CORTEVA agriscience

Summary

- Corteva is actively developing and tracking measurable sustainability parameters throughout Discovery & Development
  - Guide Research
  - Make Decisions
  - Drive Holistic Process Improvements
  - Meaningful in R&D and manufacturing (YMMV)
    - Make sustainability personal!

CORTEVA agriscience

Acknowledgments Corteva R&D Sustainability Team:

- Ashish Batra
- Jessica Herrick
- Dan Kaiser
- Daniel Valco
- John Rohanna
- Greg Whiteker (retired)

CORTEVA agriscience 23

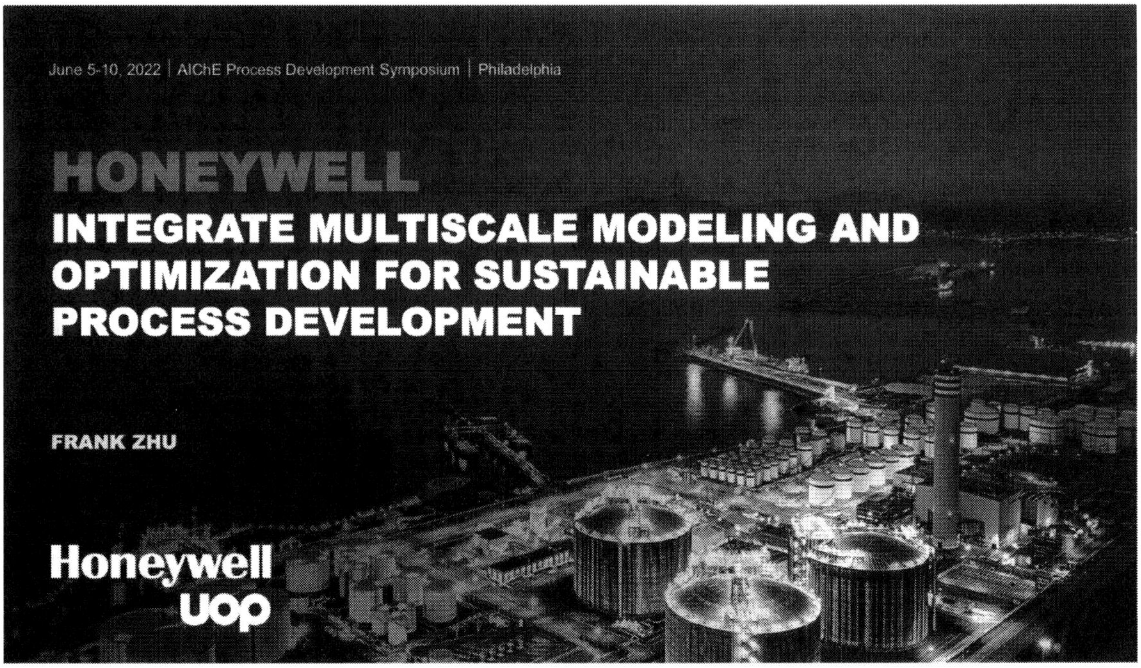
June 5-10, 2022 | AIChE Process Development Symposium | Philadelphia
HONEYWELL
INTEGRATE MULTISCALE MODELING AND OPTIMIZATION FOR SUSTAINABLE PROCESS DEVELOPMENT
FRANK ZHU
Honeywell
UOP

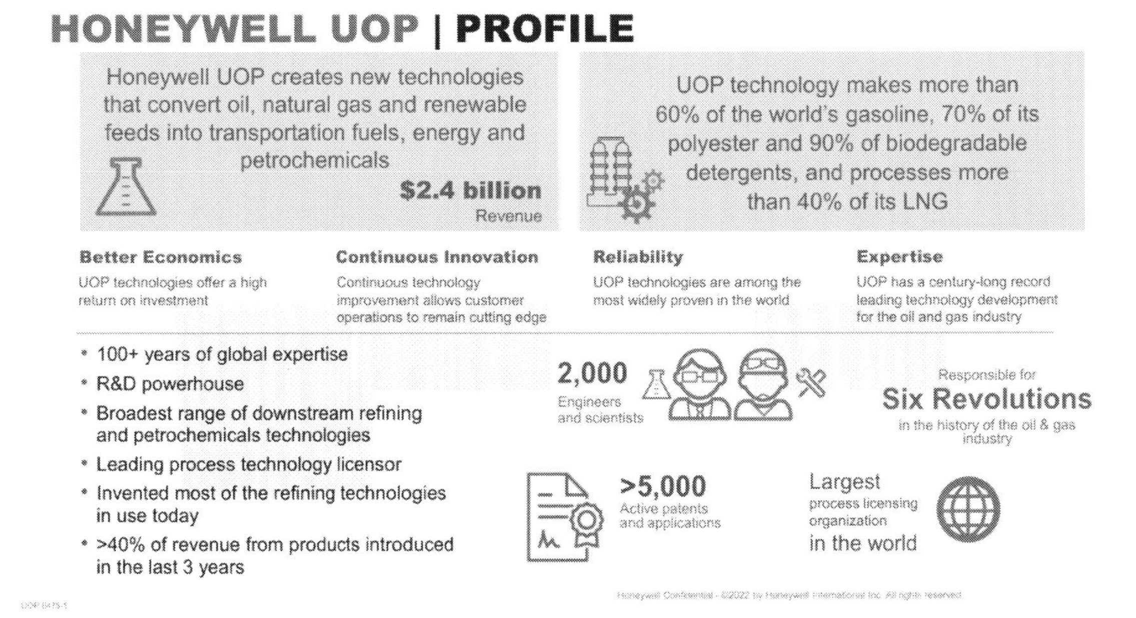
HONEYWELL UOP | PROFILE
Honeywell UOP creates new technologies that convert oil, natural gas and renewable feeds into transportation fuels, energy and petrochemicals
$2.4 billion
Revenue
UOP technology makes more than 60% of the world's gasoline, 70% of its polyester and 90% of biodegradable detergents, and processes more than 40% of its LNG
Better Economics
UOP technologies offer a high return on investment
Continuous Innovation
Continuous technology improvement allows customer operations to remain cutting edge
Reliability
UOP technologies are among the most widely proven in the world
Expertise
UOP has a century-long record leading technology development for the oil and gas industry
• 100+ years of global expertise
• R&D powerhouse
• Broadest range of downstream refining and petrochemicals technologies
• Leading process technology licensor
• Invented most of the refining technologies in use today
• >40% of revenue from products introduced in the last 3 years
2,000
Engineers and scientists
Responsible for
Six Revolutions
in the history of the oil & gas industry
>5,000
Active patents and applications
Largest
process licensing organization
in the world
Honeywell Confidential - ©2022 by Honeywell International Inc. All rights reserved.

Key Questions

- How to speed-up development cycle from discovery to commercialization for a new process?
- How to optimize reaction pathways based on economic and environmental performance?
- How to conduct process scale-up effectively?
- How to provide feedback from later design stages back to earlier stages to minimize rework?

Honeywell
UOP

Outline

- Current Practice for Process Development
- Multiscale modeling and optimization
- The new methodology for codesign of product & process
- The value by the new method
- Concluding remarks

Honeywell
UOP

4

Current Practice of Product / Process Development Phases

Macroscopic analysis:

- Reaction pathways based on known mechanism and catalyst
- Existing technology is used as the basis for selection
- Synthesis is done via rules of thumbs & in-house experience
- Optimization focuses on local systems on stand-alone basis

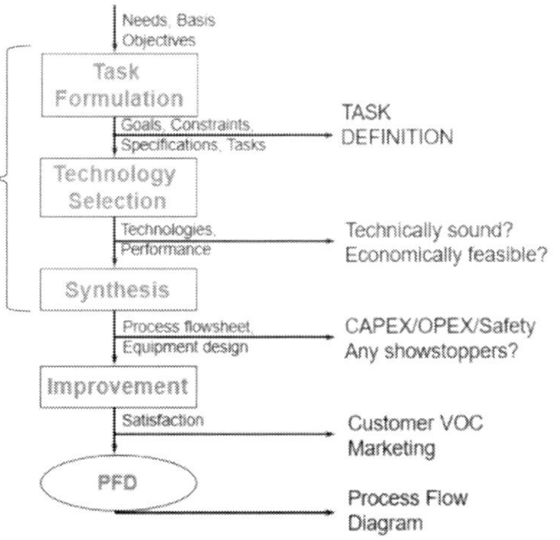

Isolation of Product & Process Design / Macro-Analysis Oriented

5

Novel Methodology for Co-design of Product and Process

Microscopic molecular analysis:

1. **Molecular modeling for optimizing reaction pathways**
2. **Quantum mechanism identifies desired properties for catalyst**
3. **Integrating MKM with CFD**
4. **Limiting-condition analysis identifies critical factors and ideas**

Macroscopic economic optimization:

5. **Superstructure optimization of technology options for overcoming limiting conditions**

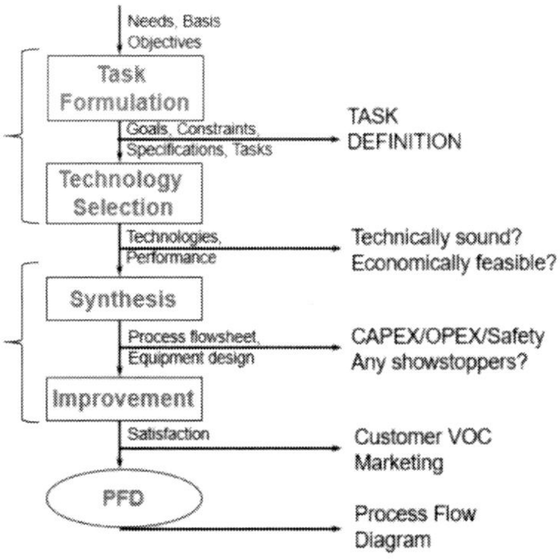

Co-development = Co-design (Chemistry + Catalyst + Reaction + Separation)

Goals of the New Method for Process Development

- Reduce process development cycle via integrating multiscale models to explore synergy between different development stages and minimize time needed for development efforts
- Maximize processing efficiency: better use of raw materials, higher yields, less energy/water/waste/emissions with minimum capital and operating costs.
- Integrating different disciplines including quantum chemistry, reaction chemistry, reaction kinetic, material science, and engineering
- The end-game goal is to generate breakthrough ideas and novel process concepts for grassroots and revamp

Honeywell
UOP

Outline

- Current Practice for Process Development
- **Multiscale modeling and optimization**
- The new methodology for codesign of product & process
- The value by the new method
- Concluding remarks

Honeywell
UOP

What is Multiscale?

-- From Molecular to Particles to Equipment to Process
-- From Lab to Pilot Plant to Commercialization

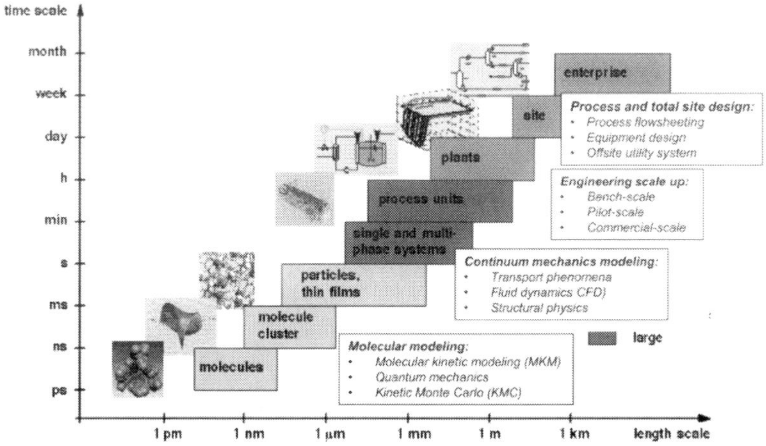

Physics vs. Length and Time Scales

Honeywell UOP

Molecular Modeling Identifies Multiple Reaction Routes

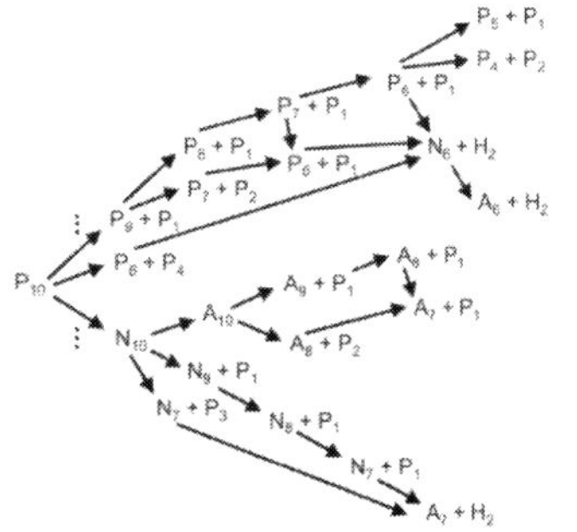

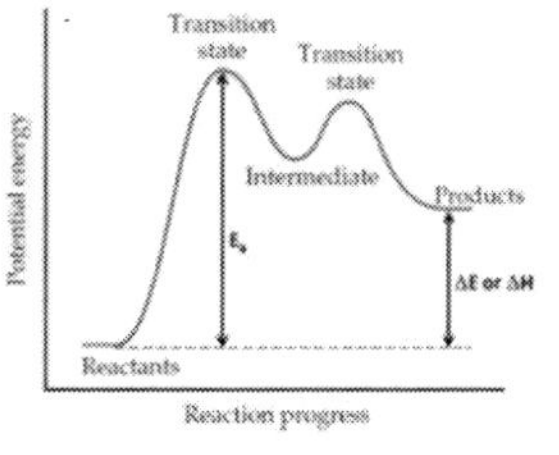

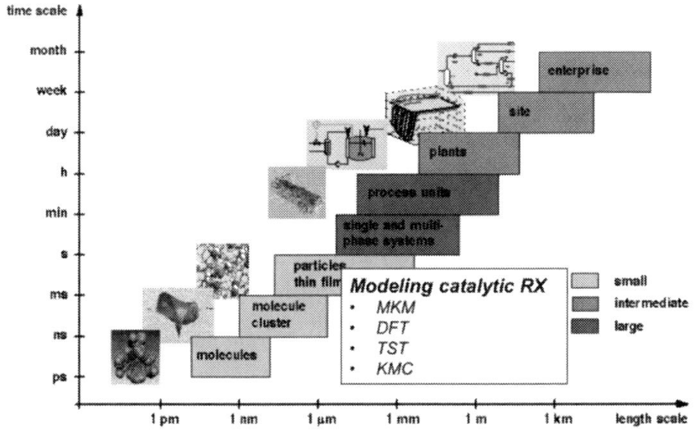

Honeywell UOP

Molecular Modeling Integrating with CFD Flow Modeling

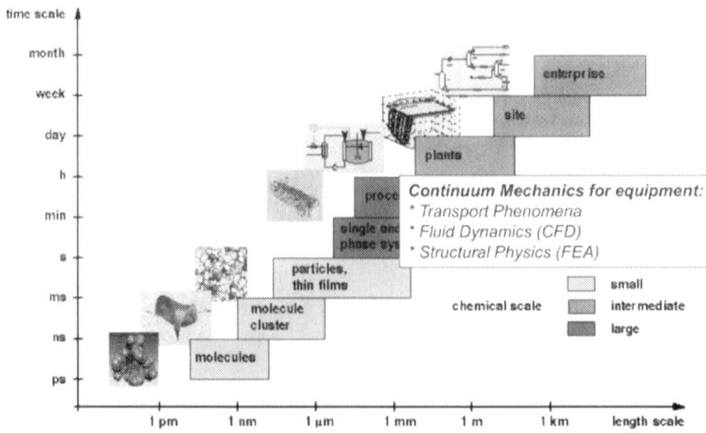

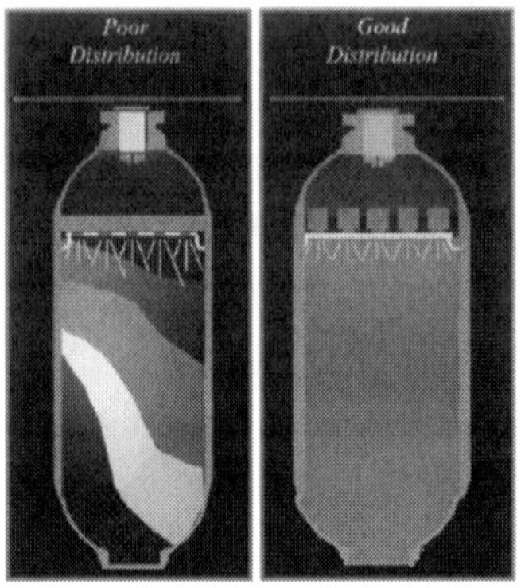

Integrating MKM and CFD intends to minimize scale-up effort

Process Superstructure Captures Selected Technology Options

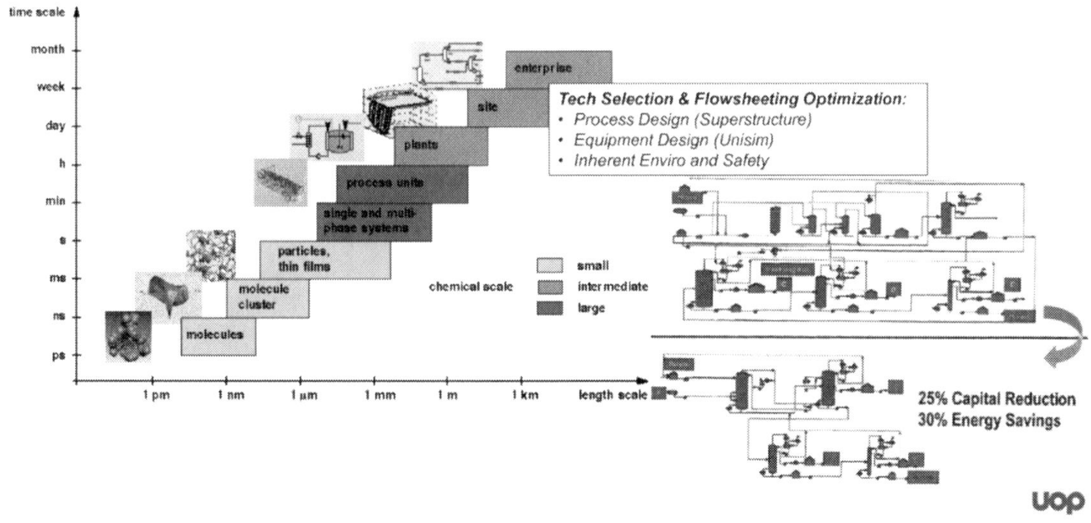

UOP

12

6-Pillar Based Approach for Co-Design

Pillar 1: Reaction Chemistry Identification

- Feed and catalyst characterizations
- Reaction mechanism determination
- Reaction network generation

Pillar 2: Molecular-based Mechanistic Kinetic Modeling

- Reaction kinetics and rate
- Catalyst Structure-Activity relationships
- Catalyst morphology

Pillar 3: Continuum Mechanics Modeling

- Reactor type vs. internal
- RX conversion vs residence time
- Flow patterns vs mixing vs temperature profile

Pillar 4: Process Simulation & Integration

- Determine initial process flowsheeting
- Identify hybrid systems
- Determine multiple options for technology selection

Pillar 5 Superstructure Optimization

- The superstructure consisting of all promising options
- Determine technology selection

Pillar 6: Limiting Condition Analysis

- Define the critical factors for improvements
- Identify options for overcoming limiting conditions

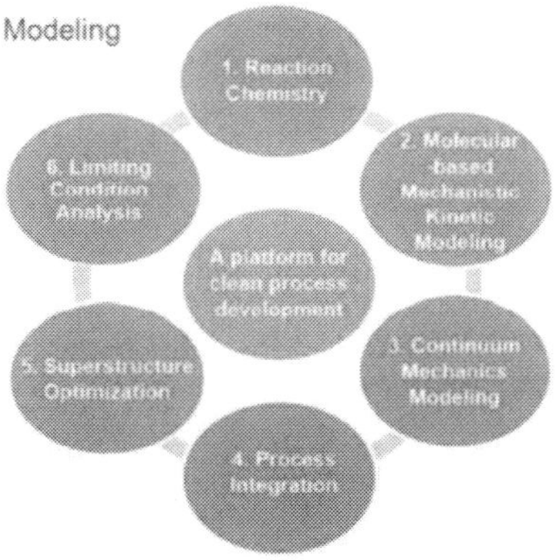

Use molecular analysis to guide catalyst design & Use LCA to provide feedback

13

New Co-Design Methodology

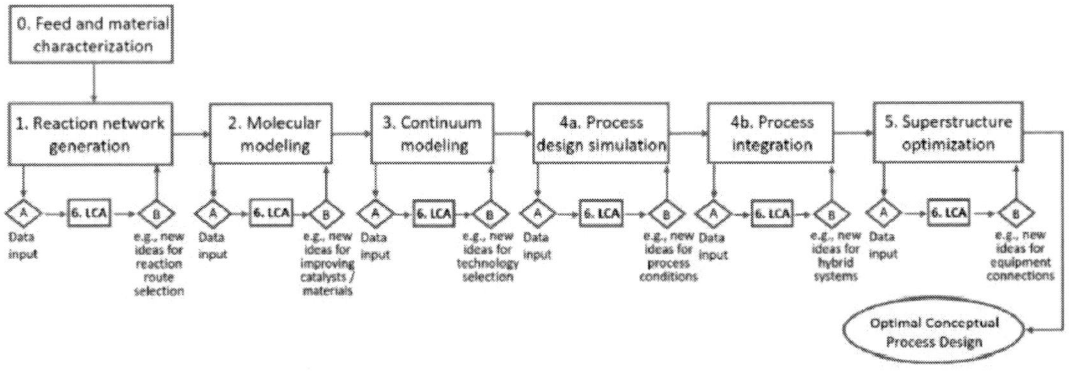

Frank Zhu, LJ Xu (2022), Integrating Multiscale Modeling and Optimization for Sustainable Process Development, Vol. 254, 117619, CES.

Honeywell UOP

UOP 8299-13

14

Outline

- Current Practice for Process Development
- Multiscale modeling and optimization
- The new methodology for codesign of product & process
- **The value by the new method**
- Concluding remarks

Honeywell
UOP

15

The Value from the new method for Process Development

❑ Primary value via molecular modeling and analysis

- Explore synergy between quantum chemistry, reaction kinetics, bio-chemistry and chemical engineering
- Effective ways for overcoming reaction limiting steps
- Novel processing concepts
- Efficient conversion of feedstocks to required products

❑ Secondary value via superstructure optimization

- Determine the best combination of technology options
- Efficient process design with lower Capex/Opex
- Integrated batch/continuous processes
- Hybrid systems design
- Process and equipment intensification
- Flexibility, reliability, safety and environment

Can be applied for both grassroots and revamp

From Crude to Petrochemicals

"All design starts with a perceived need. In the design of a chemical process, the need is the public need for the product, creating a commercial opportunity, as foreseen by the sales and marketing organization."

Towler, G., & Sinnott, R. (2008). *Chemical Engineering Design*. Elsevier/Butterworth-Heinemann. Chapter 1, p. 3.

Honeywell UOP

Petrochemicals Show Consistently Strong Growth

Global Petrochemical Mix

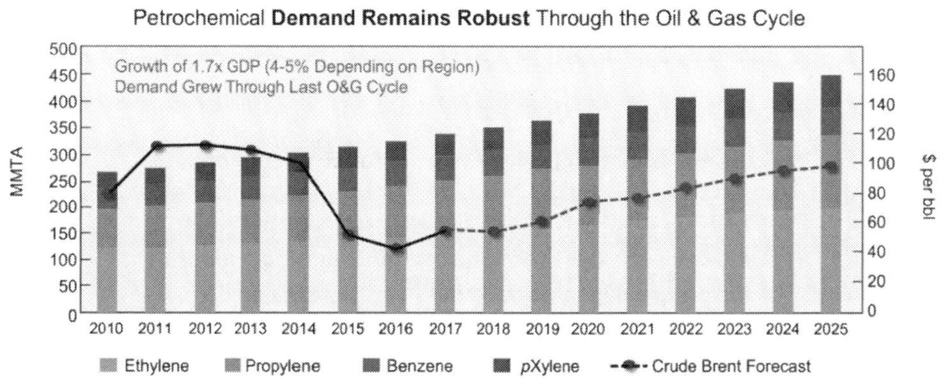

Honeywell UOP

18

Petrochemicals Value Add

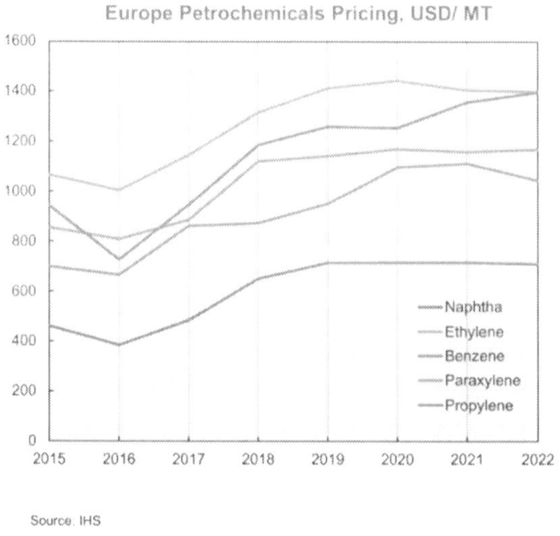

Source: IHS

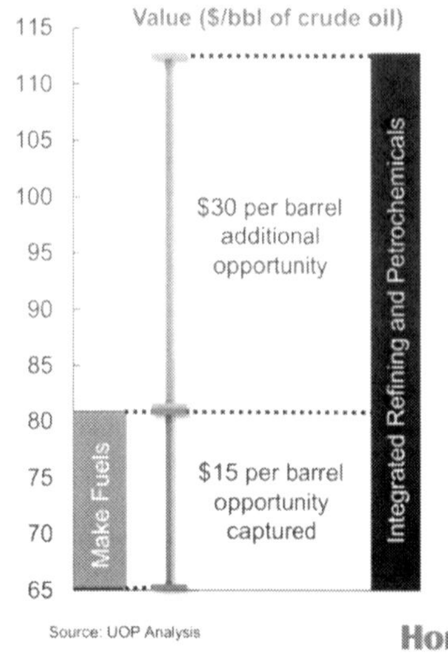

Source: UOP Analysis

Honeywell UOP

19

Reconfiguring Towards Petrochemicals

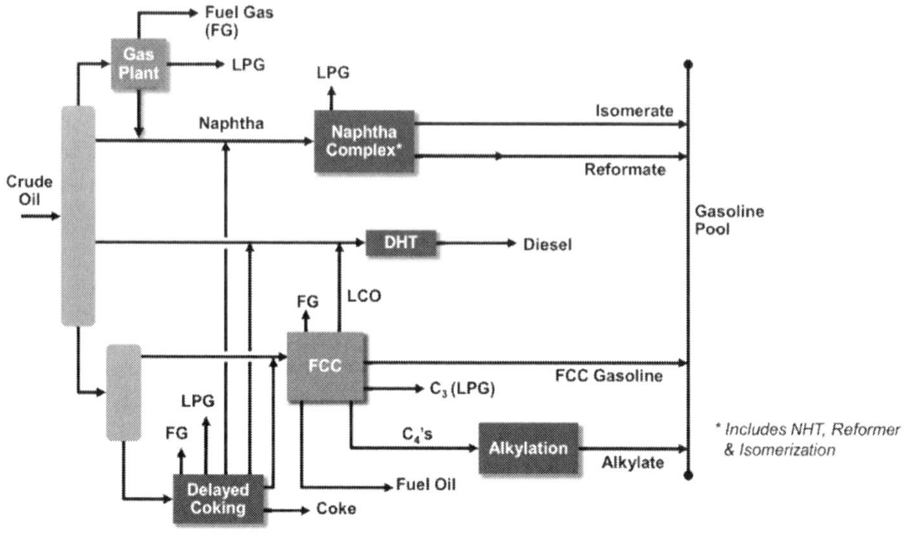

* Includes NHT, Reformer & Isomerization

Honeywell UOP

20

Reconfiguring Towards Petrochemicals

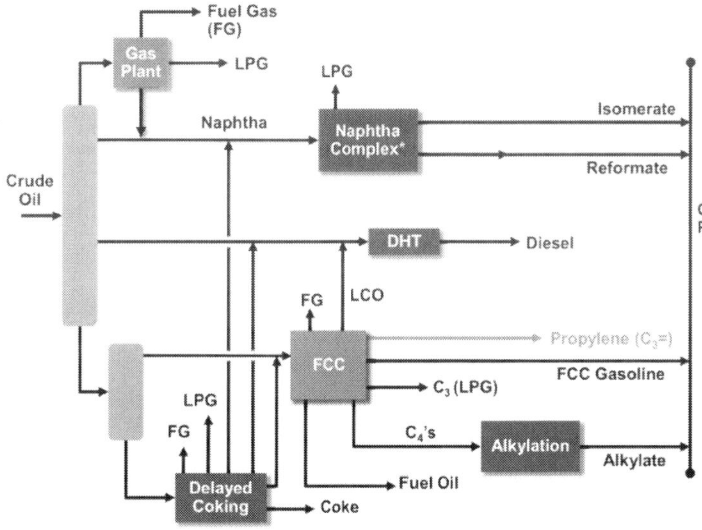

* Includes NHT, Reformer & Isomerization

Honeywell UOP

UOP 8399-20

21

Reconfiguring Towards Petrochemicals

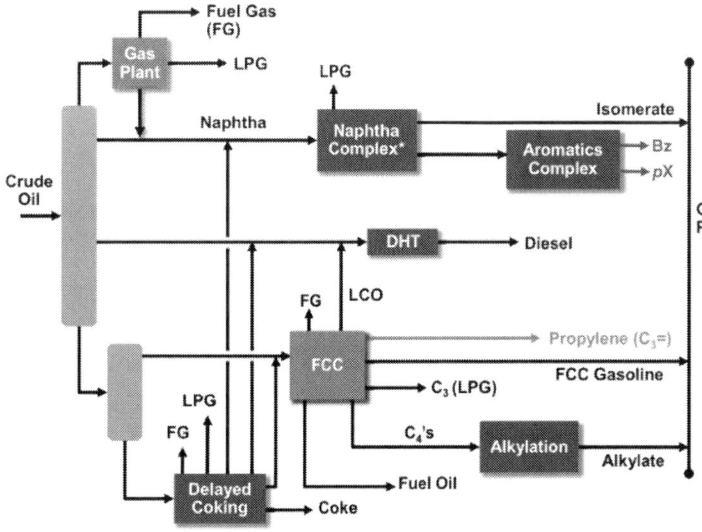

* Includes NHT, Reformer & Isomerization

Honeywell UOP

UOP 8399-21

22

Reconfiguring Towards Petrochemicals

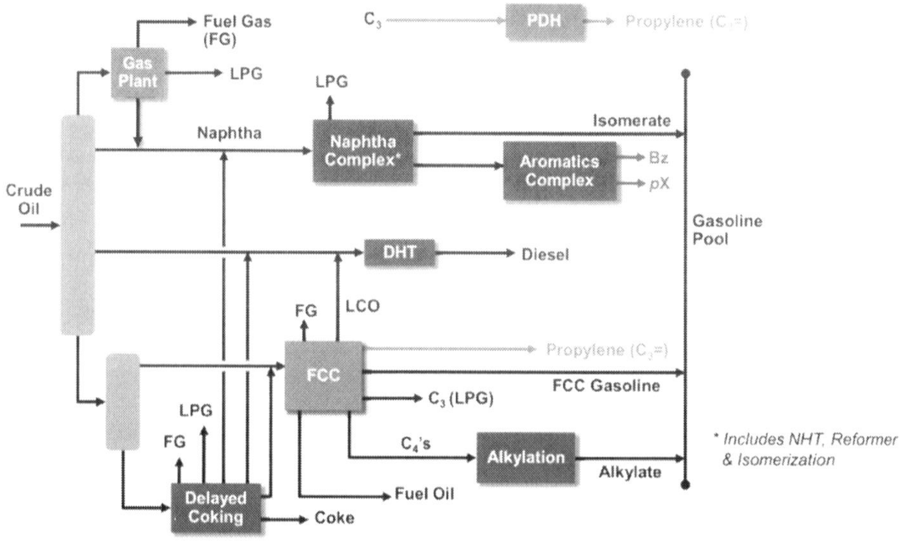

* Includes NHT, Reformer & Isomerization

Honeywell UOP

23

Reconfiguring Towards Petrochemicals

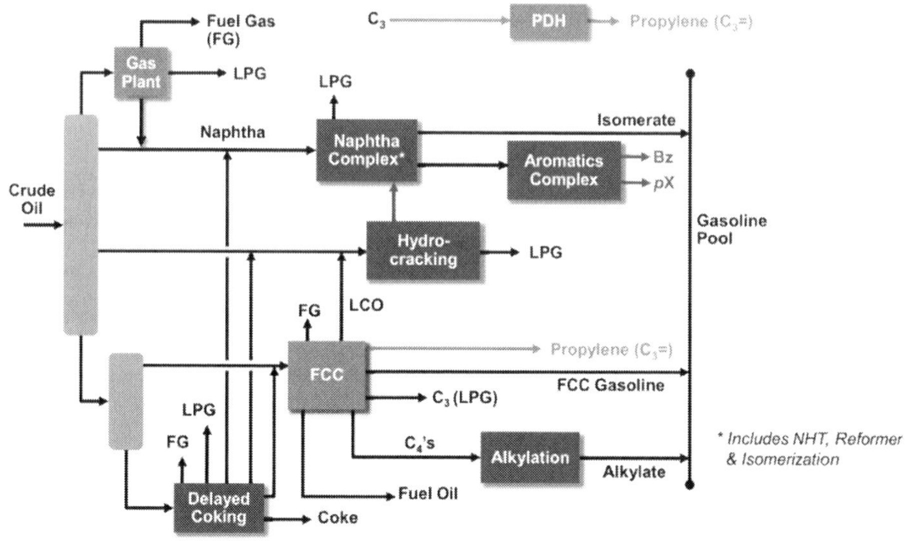

* Includes NHT, Reformer & Isomerization

Honeywell UOP

Reconfiguring Towards Petrochemicals

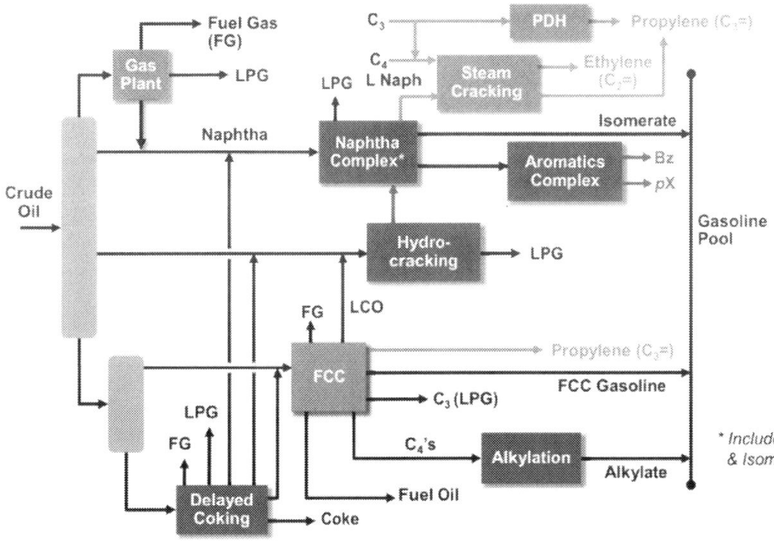

* Includes NHT, Reformer & Isomerization

Honeywell UOP

SUSTAINABLE TECHNOLOGY SOLUTIONS

A New Business Unit for Honeywell

RENEWABLES FUEL

Reducing Diesel and Jet GHG Emissions >80%*

UOP Ecofining™ can deliver **outsized profits** plus **significant reduction in GHG emissions**

PLASTICS CIRCULARITY

Removing Oil & Gas Extraction from Plastic Production

Honeywell Upcyle Plastics Recycling **increases waste plastic circularity** if collection and sorting is improved

ENERGY STORAGE

Driving toward Zero Emissions Power

Honeywell Renewable Energy Solutions help **optimize energy** with **energy storage** and remote operations

HYDROGEN

Low Emissions Fuel for Hard to Decarbonize Applications

Honeywell UOP provides **efficient, tailored CO_2 capture** for hydrogen production

CARBON CAPTURE

Reducing CO_2 Emission

Ready-now technology to significantly **reduce CO_2 emissions** from energy and industrial point sources

Honeywell UOP

APPLY NEW METHOD FOR SUSTAINABLE PROCESS DEVELOPMENT

Molecular analysis can unlock the myth of value chain in catalyst/process development while superstructure optimization can determine the fit-for-purpose technologies and connections among them (configuration) with the objective of achieving optimal economical and environmental performance

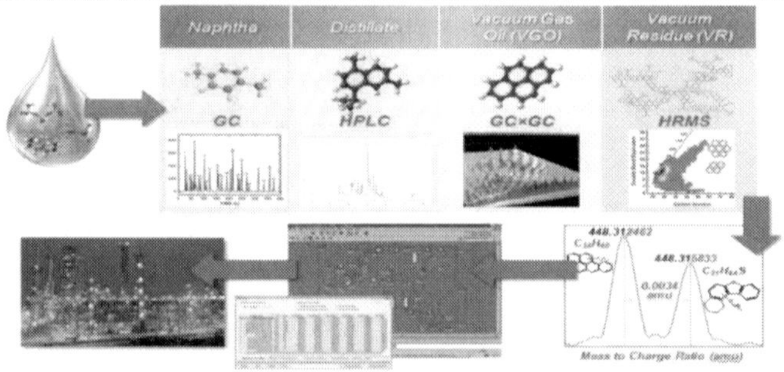

Honeywell UOP

Forward Looking Statements

This presentation contains certain statements that may be deemed "forward-looking statements" within the meaning of Section 21E of the Securities Exchange Act of 1934. All statements, other than statements of historical fact, that address activities, events or developments that we or our management intends, expects, projects, believes or anticipates will or may occur in the future are forward-looking statements. Such statements are based upon certain assumptions and assessments made by our management in light of their experience and their perception of historical trends, current economic and industry conditions, expected future developments and other factors they believe to be appropriate. The forward-looking statements included in this presentation are also subject to a number of material risks and uncertainties, including but not limited to economic, competitive, governmental, technological, and COVID-19 public health factors affecting our operations, markets, products, services and prices. Such forward-looking statements are not guarantees of future performance, and actual results, and other developments, including the potential impact of the COVID-19 pandemic, and business decisions may differ from those envisaged by such forward-looking statements. Any forward-looking plans described herein are not final and may be modified or abandoned at any time. We identify the principal risks and uncertainties that affect our performance in our Form 10-K and other filings with the Securities and Exchange Commission.

Honeywell UOP

26
follow us
Honeywell
UOP
https://www.uop.com/
https://twitter.com/HoneywellUOP
Customer Portal
https://www.accessuop.com/
You Tube
https://www.youtube.com/user/Honeywell
https://www.linkedin.com/company/uop
Honeywell
UOP

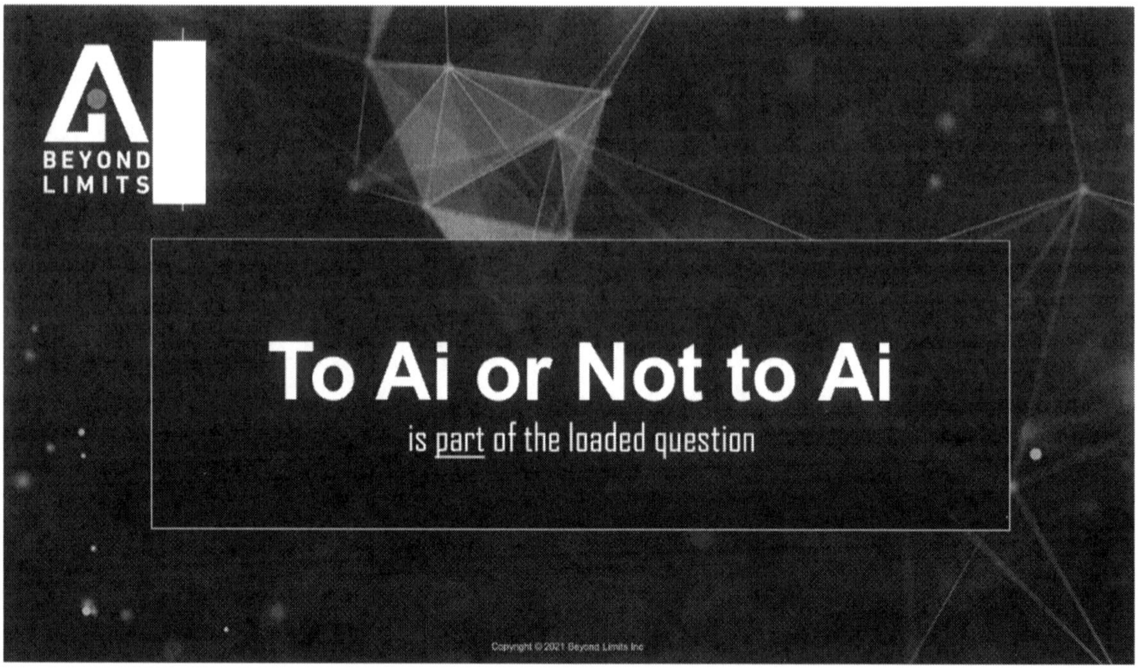
BEYOND
LIMITS
To Ai or Not to Ai
is part of the loaded question
Copyright © 2021 Beyond Limits Inc

About the Speaker
Chris Clayton
Head of Customer Success
Beyond Limits
PR@beyond.ai
cclayton@beyond.ai
Expectations for this talk:
• Generalist: AI, Big Data Analytics, Solution Architecture, UX, UI, Design Thinking
• Roles: Sales, Bus Dev, Functional Product Management, Client Partner/Customer Success, Professional Services/Delivery

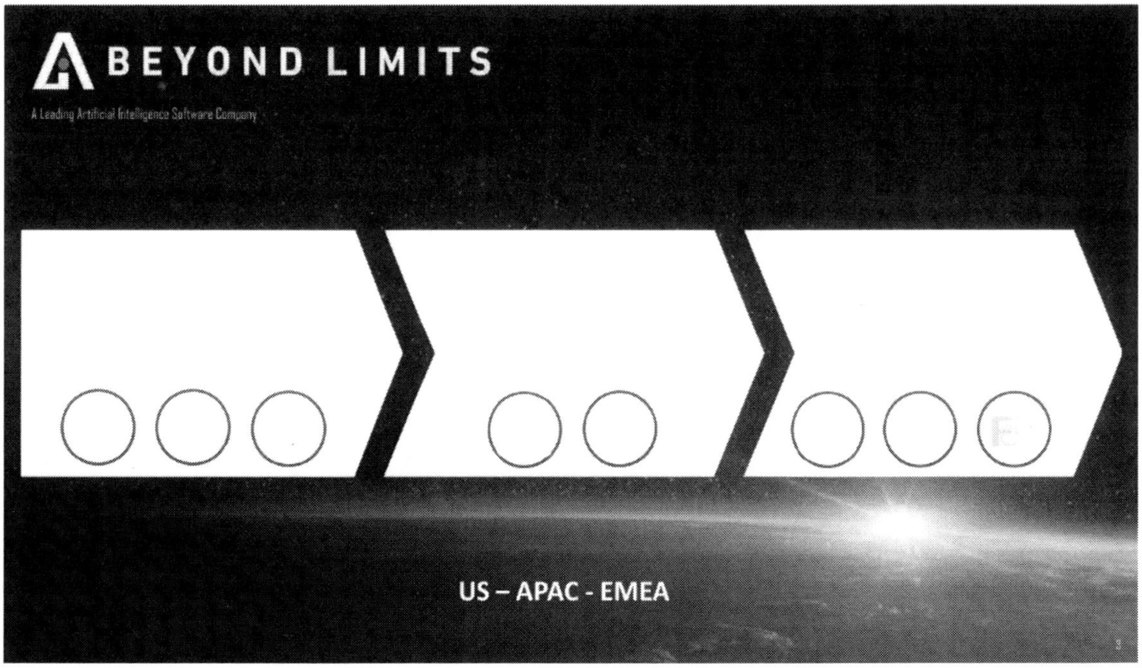
BEYOND LIMITS
A Leading Artificial Intelligence Software Company
US – APAC - EMEA

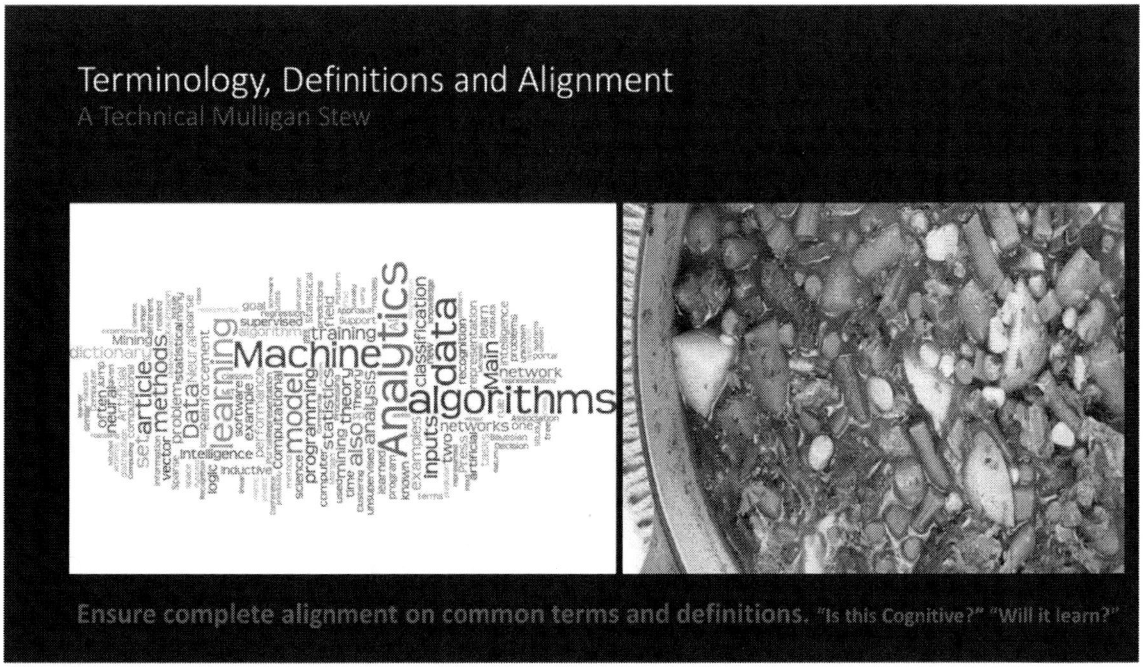
Terminology, Definitions and Alignment
A Technical Mulligan Stew
Machine
Analytics
algorithms
learning
data
Ensure complete alignment on common terms and definitions. "Is this Cognitive?" "Will it learn?"

Managing Unrealistic Expectations

Context is Everything

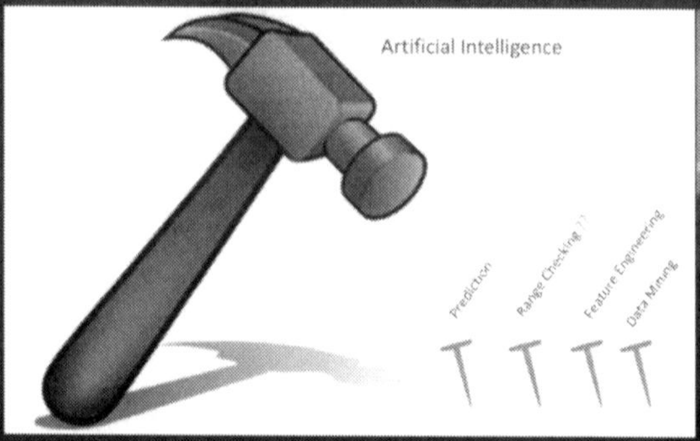

Data, Decision Making & AI Approaches

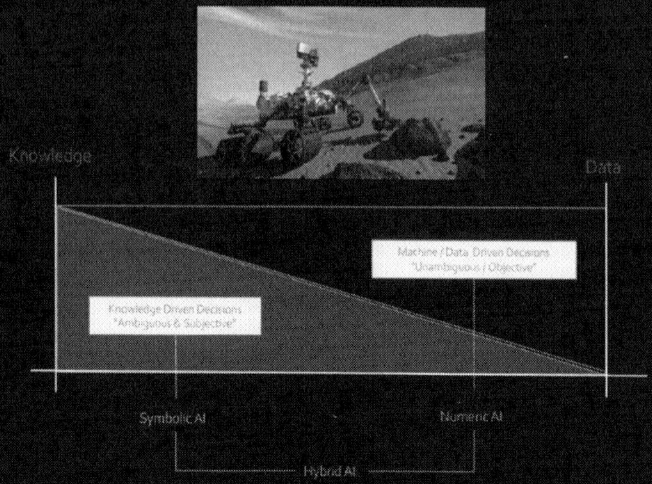

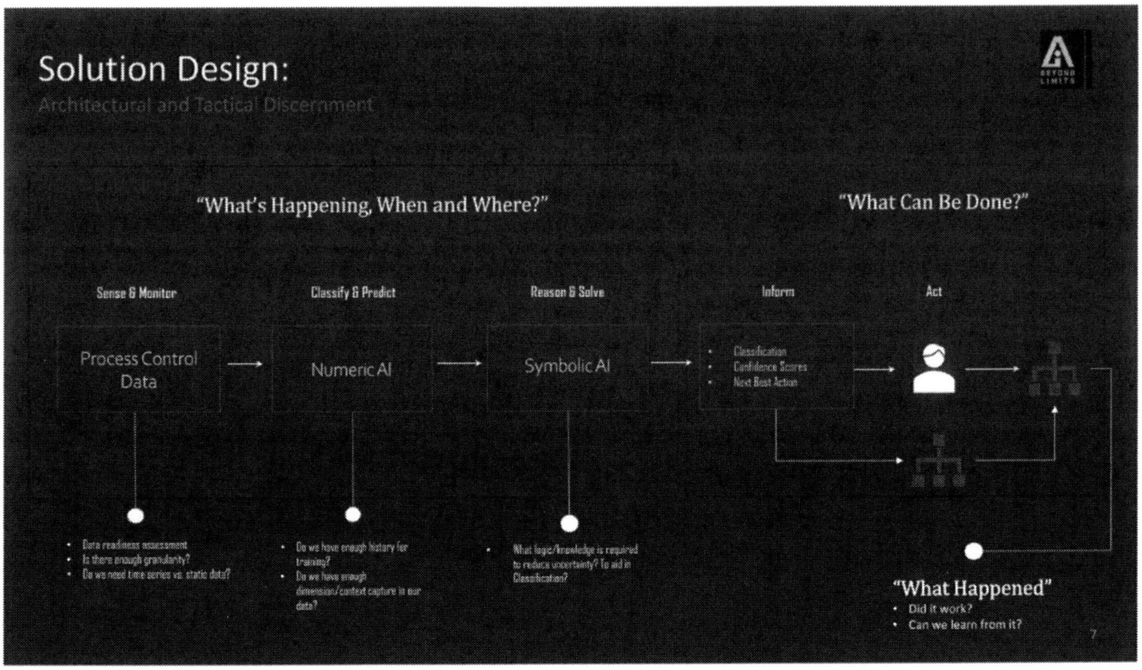

Strategy and Planning Tools

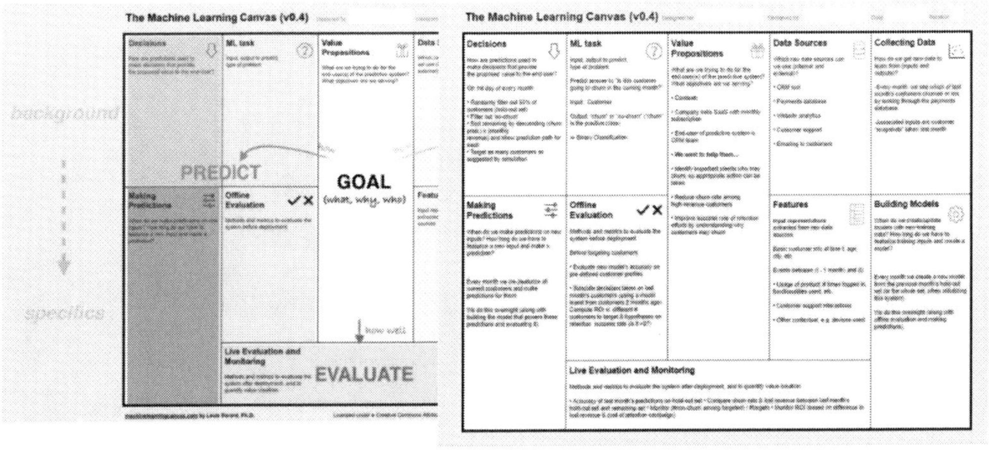

https://www.ownml.co/machine-learning-canvas

Context

- Oil Refinery Optimization
- 450K B/day
- Feedstock Variability
- 200 + Objectives

Challenges

- Close the Profitability Gap
- Time to Decision..... Time is money
- Global Situational Awareness

Problem Definition & Understanding

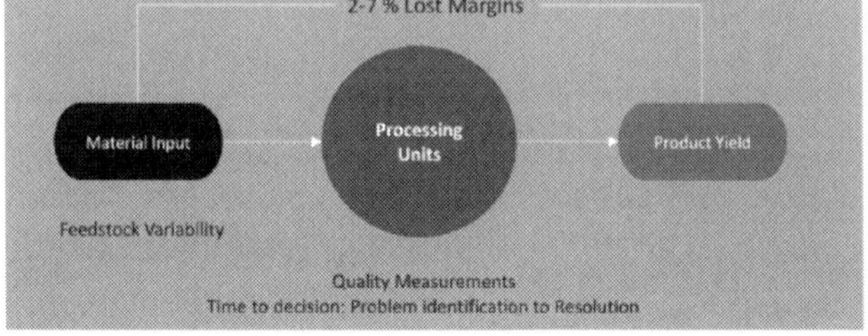

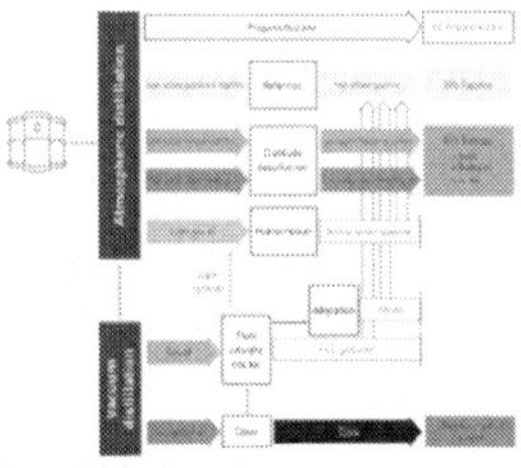

Example of a Complex Refinery

Solution Design:

Align on Goals, Methods & Expectations

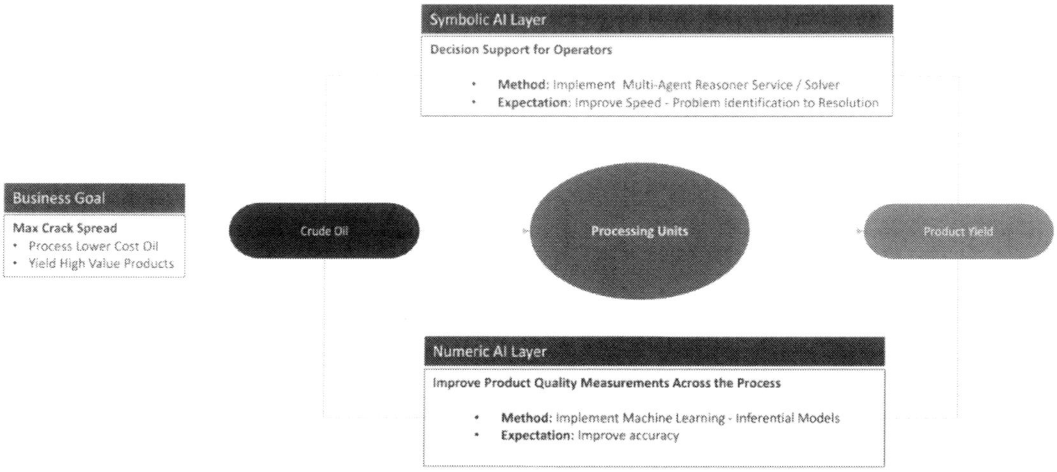

Symbolic AI: Multi-Agent Framework

Evaluate, Reason and Solve

- Modular agents with local configurations
- Agents can find either local or non-local solutions
- Global resolution ensured by multi-agent interactions

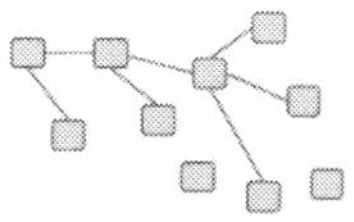

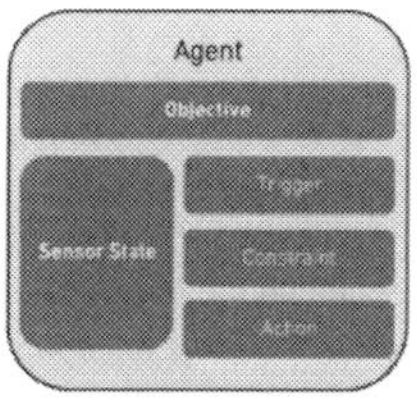

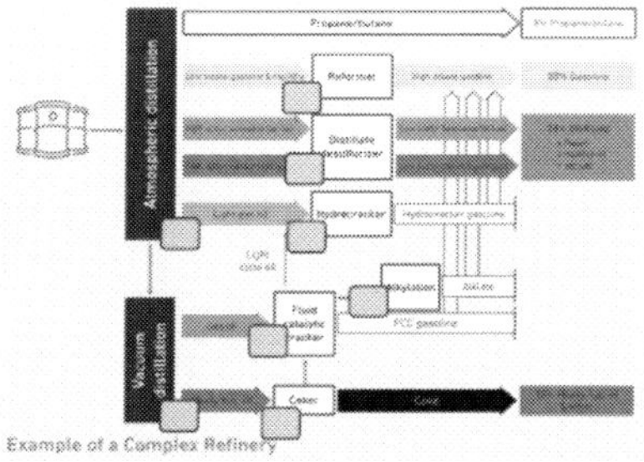

Example of a Complex Refinery

The Refinery Advisor

Accelerate Digital Innovation & Optimize Human Performance

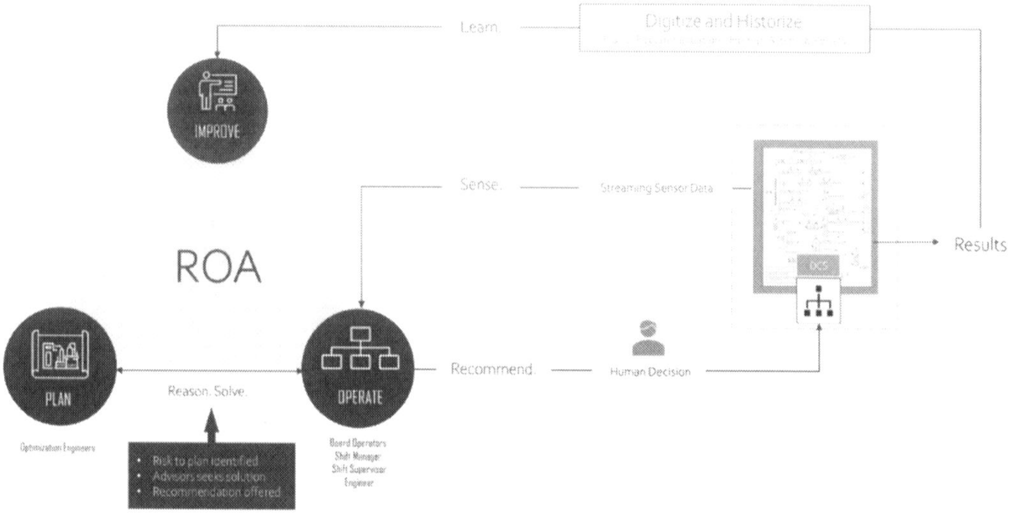

Symbolic Reasoning Example

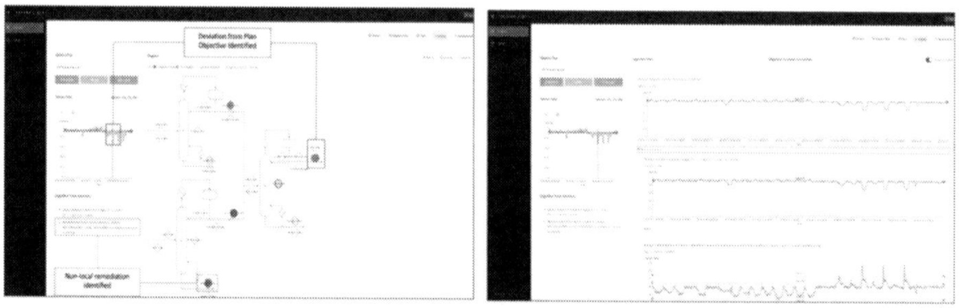

USE CASE	SENSE & MONITOR	REASON & SOLVE	ACT	Value
Production Optimization	**Units/Systems / Subsystems**	Identify off plan behavior	**Advisor provides resolution recommendations with explanation of reasoning.**	Uniform application of safety, risk and operational best practices Improve Profitability
	Temperature	Industry and Expert Knowledge		
	Pressure	First Principles or Physics Models		
	Lab Assays (feedstock characteristics)	Interpretate & apply best practices		Embed operational excellence and continuous learning into the workflow.
	Humidity	Apply Reasoning/Logic		
	External Temperature			

14

Numeric Layer

ML Based Product Quality Predictions

Existing Inferentials	Attributes	ML Based Inferentials
Limited to 2 Variables	Data Inputs	Using all potential variables (40 +)
Domain Knowledge/Experience	Feature Selection	Statistical methods
Linear	Model Type	Linear; Non-linear
Set & forget	Model Maintenance	Continuous re-training & updating
Simple and easy to implement	Value	• Typically generates higher accuracy • Enables feature importance analysis • Reduce dependency on labs and analyzers

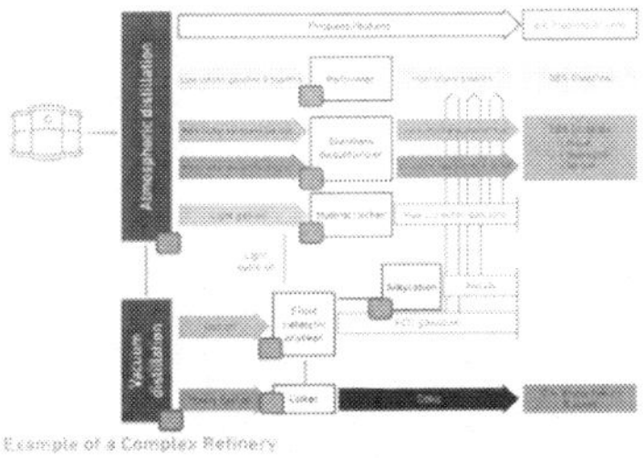

Example of a Complex Refinery

BEYOND LIMITS

Q & A

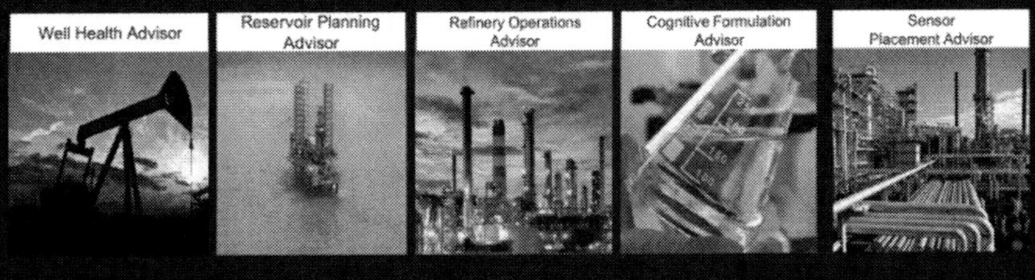

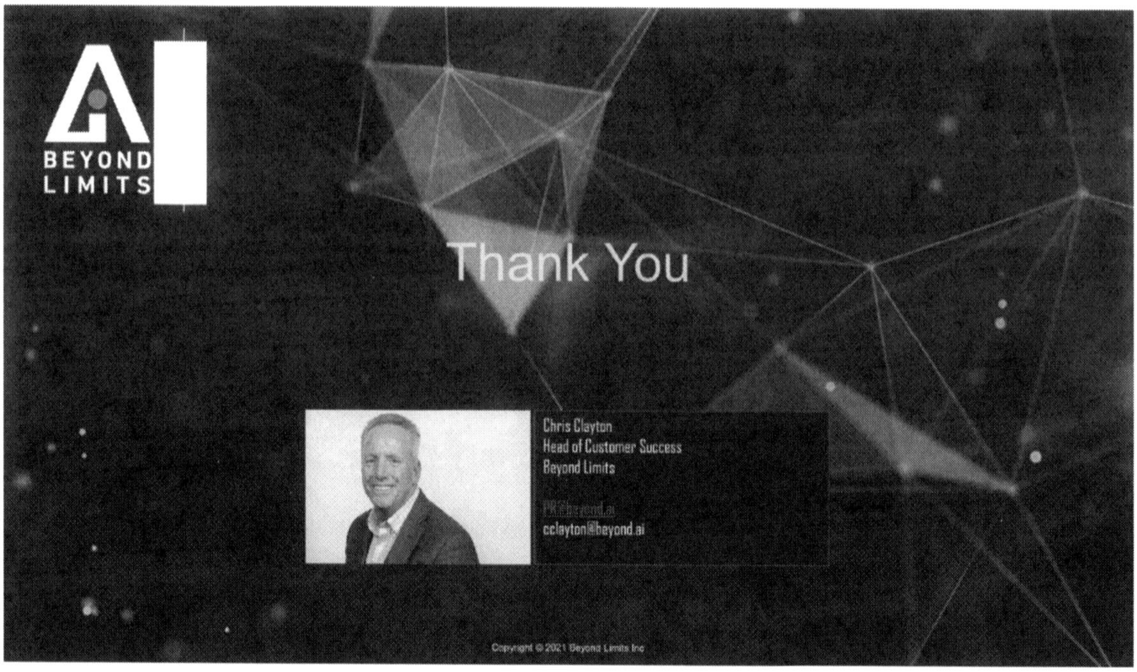
BEYOND
LIMITS
Thank You
Chris Clayton
Head of Customer Success
Beyond Limits
PR@beyond.ai
cclayton@beyond.ai
Copyright © 2021 Beyond Limits Inc

pds | PROCESS DEVELOPMENT SYMPOSIUM 2022

JUNE 7-9, 2022
PHILADELPHIA, PA

Panel Discussion / Interactive Session

pds | PROCESS DEVELOPMENT SYMPOSIUM 2022
JUNE 7-9, 2022 • PHILADELPHIA, PA

In the next 90 minutes…

- Quick demographic snapshot
- Panel Discussion
- Small Group Discussion
- Small Group Feedback

- Output from this session will be included in the proceedings

Organized by AIChE® and the Process Development Division

PROCESS DEVELOPMENT SYMPOSIUM 2022
JUNE 7-9, 2022 • PHILADELPHIA, PA

Your facilitators

- Aaron Sarafinas
  - Independent consultant since June 2018
    - with over 35 years of experience
    - being the bridge between the lab and the plant
    - at Rohm and Haas and The Dow Chemical Company
  - Helping my clients
    - achieve faster process development, troubleshooting, and scaling
    - enabling my clients to efficiently understand how mixing matters to their process and what they can do about it
  - Delivering technical training, including
    - "Scaling and Piloting for the Process Industries" for CfPA
    - "Applied Mixing Technology for Process Development: Scaling Down to Scale Up" for Scientific Update
  - And I've facilitated these interactive sessions many times since our first PDS in 2003
- Deboleena Chakraborty
  - Research Scientist in Core R&D at Dow
    - 12 years with Dow and total of 15 years of industrial R&D experience
    - Process development Engineer with expertise in providing commercialization technical support
    - Techno-economic feasibility analysis (TEA)
    - Sustainability assessments (LCA) to facilitate product development for a wide range of commercial application
  - Represent Dow at The American Center for Life Cycle Assessment (ACLCA) board
  - Been a part of PDS since 2018

Organized by AIChE® and the Process Development Division

PROCESS DEVELOPMENT SYMPOSIUM 2022
JUNE 7-9, 2022 • PHILADELPHIA, PA

Quick Demographic Snapshot

- Let's get a quick visual image of who is at the 2022 PDS
- Example: If you consider yourself an extrovert, stand up.
- We'll take a quick look at:
  - Engineer / Chemist / Other
  - Industry / Academia / Consultant
  - R&D / Engineering / Manufacturing / Business / Other
  - Management Track / Technical Ladder / Undecided
  - Years since your Bachelor degree
  - Market segment
  - What you do with AIChE

Organized by AIChE® and the Process Development Division

PROCESS DEVELOPMENT SYMPOSIUM 2022
JUNE 7-9, 2022 • PHILADELPHIA, PA

Market Segments

- "Materials" vs "Equipment" vs "Tools" vs "Services"
- "Chemicals"
  - Commodity / specialty
  - "Active Ingredients" / "Formulated Products"
- "Pharmaceuticals"
  - API's
  - Final Dosage Formulations
  - Medical Devices

Organized by AIChE® and the Process Development Division

PROCESS DEVELOPMENT SYMPOSIUM 2022
JUNE 7-9, 2022 • PHILADELPHIA, PA

What I do with AIChE?
I'm attending this conference…1st time?

- member
- member of a division (member of PDD?)
- member of a local section (and "active"?)
- look at CEP
- read (at least skim) weekly AIChE SmartBrief or monthly AIChExchange
- use AIChE social media
  - Engage Discussion Central
  - ChEnected
- taken a course or webinar through AIChE Academy
- ever attended a monthly "virtual local section" presentation?

Take advantage of what the AIChE has to offer…

Organized by AIChE® and the Process Development Division

PROCESS DEVELOPMENT SYMPOSIUM 2022
JUNE 7-9, 2022 • PHILADELPHIA, PA

In the next 80 minutes…

- Quick demographic snapshot
- Panel Discussion
  - Quick reintroduction
  - Discuss some questions
  - Lead-in to small group discussions
- Small Group Discussion
- Small Group Feedback

Output from this session will be included in the proceedings

Organized by AIChE® and the Process Development Division

PROCESS DEVELOPMENT SYMPOSIUM 2022
JUNE 7-9, 2022 • PHILADELPHIA, PA

So far, we've heard…

Advancing Process Development: Design, Innovation, and a Circular Economy

- Keynote
  - Dominic T. Clausi, *ExxonMobil*
- Driving Innovation in Process Development
  - Martin Johnson, *Eli Lilly*
  - Cheryl Teich, *Teich Process Development LLC*
  - Darren Youngs, *Lanzatech*
- Designing Sustainable Processes
  - Abraham Schuitman, *Corteva Agriscience*
  - Benjamin Cohen, *Bristol Myers Squibb*
  - Santiago Faucher, *Ecomaterials Inc.*

Organized by AIChE® and the Process Development Division

PROCESS DEVELOPMENT SYMPOSIUM 2022
JUNE 7-9, 2022 • PHILADELPHIA, PA

Advancing Process Development: Design, Innovation, and a Circular Economy

So far, we've heard...

- Process Data Collection, Analysis, and Modeling for Lab, Pilot, and Production Plants
  - Travis Armiger, *Merck*
  - Xin (Frank) Zhu, *Honeywell*
  - Chris Clayton, *Beyond Limits*

Organized by AIChE® and the Process Development Division

PROCESS DEVELOPMENT SYMPOSIUM 2022
JUNE 7-9, 2022 • PHILADELPHIA, PA

Some words of wisdom from today...

- "current technology is insufficient to navigate energy transition"
- "integrating affordable and sustainable solutions"
- "there will be many solutions...not just one solution"
- "the world is going to need negative emissions technology...we know it's doable"

Organized by AIChE® and the Process Development Division

PROCESS DEVELOPMENT SYMPOSIUM 2022
JUNE 7-9, 2022 • PHILADELPHIA, PA

More words of wisdom from today…

- Good (properly placed) HPLC data are boring, but deviations tell much about process performance in real-ish time
- "the goal is commercial success"
- "decreasing time to profit without sacrificing understanding"
- overlay reaction and separation process understanding with modular knowledge base

Organized by AIChE® and the Process Development Division

PROCESS DEVELOPMENT SYMPOSIUM 2022
JUNE 7-9, 2022 • PHILADELPHIA, PA

More words of wisdom from today…

- "predictions are simply extrapolations of the past...innovations expands the art of what is possible"
- "biological processes – self-regenerating once inhibitor is removed" as opposed to thermal-catalytic poisoning compromised performance until regeneration or replacement
- 1MM tonnes per year sequestered into products

Organized by AIChE® and the Process Development Division

PDS | PROCESS DEVELOPMENT SYMPOSIUM 2022
JUNE 7-9, 2022 • PHILADELPHIA, PA

More words of wisdom from today…

- "keep the endpoint in mind…make sustainability personal"
- "Chemists and engineers design, optimize and implement commercial manufacturing processes with the goals of quality, robustness, cost efficient and sustainability"
- "collecting greenness metrics is key to making decisions, but is challenging given the nature of modern API synthesis. New tools can help bridge the gap."
- "I love solids…have fun, sustainability is in your nature"

Organized by AIChE® and the Process Development Division

PDS | PROCESS DEVELOPMENT SYMPOSIUM 2022
JUNE 7-9, 2022 • PHILADELPHIA, PA

More words of wisdom from today…

- Flush tees for start-up, shutdown, troubleshooting…filters and data collection throughout…and solvent runs…value of data-rich expts
- "comparison across reactor scales away from target conditions can help inform future process characterization experiments"
- "Frank, can you save 50% of my development time through modeling?"
- "What is limiting?" "All design starts with a perceived need"
- "Ensure complete alignment on common terms and definitions"

Organized by AIChE® and the Process Development Division

Commercial Success and Sustainability

- "The goal is commercial success" … and *sustainability*…
  - "The world is going to need negative emissions technology"
  - "Every waste is a resource"
  - CCS if often described as pushing CO2 underground for enhanced oil recovery. Given that LanzaTech is turning CO2 into products using biology, how is underground sequestration of CO2 not a waste of value?
  - Where (efficiency and decarbonization) will we cross the 2C curve and when?
  - What does the LCA look like for Direct Air Capture?
  - What about infrastructure needs for the circular economy? Skills needs?

pds | PROCESS DEVELOPMENT SYMPOSIUM 2022
JUNE 7-9, 2022 • PHILADELPHIA, PA

Commercial Success and Sustainability

- Six pillars of route selection – how are the pillars weighted?
  - Cost of manufacturing
  - Manufacturing complexity
  - EH&S
  - IP
  - Regulatory & Biology
  - Sustainability

Implementation suggestions?

Organized by AIChE® and the Process Development Division

PROCESS DEVELOPMENT SYMPOSIUM 2022
JUNE 7-9, 2022 • PHILADELPHIA, PA

Speed from lab to commercialization

- How much experimentation can be eliminated by standardization, modeling, and AI?
- What additional individual skills and collective capabilities are needed to optimize the use of model-guided experimentation?
- How can model-guided experimentation be expanded to provide insights into sustainability? What are the challenges in modeling the different length scales and time scales involved?

Organized by AIChE® and the Process Development Division

PROCESS DEVELOPMENT SYMPOSIUM 2022
JUNE 7-9, 2022 • PHILADELPHIA, PA

In the next 50 minutes…

- Quick demographic snapshot
- Panel Discussion
- Small Group Discussion
- Small Group Feedback

Output from this session will be included in the proceedings

Organized by AIChE® and the Process Development Division

PROCESS DEVELOPMENT SYMPOSIUM 2022
JUNE 7-9, 2022 • PHILADELPHIA, PA

Assembling the small groups

- RULES
  - self-selecting into groups of about 10-12
    - based on theme questions
  - no one that you work with in your group!
    - ideally, with 10 people you don't know
- WHAT TO DO
  - introduce yourself, hand out business cards if you have them
  - select someone to take notes and report back to all
    - NOT one of today's presenters
  - Investigate your topic

Organized by AIChE® and the Process Development Division

PROCESS DEVELOPMENT SYMPOSIUM 2022
JUNE 7-9, 2022 • PHILADELPHIA, PA

Small Group Discussion Topics - 1

- Sustainability…
  - In what ways are corporate sustainability goals SMART…or not?
  - What gaps are there in the technical skills around sustainability, and how can those gaps be filled?

Organized by AIChE® and the Process Development Division

PROCESS DEVELOPMENT SYMPOSIUM 2022
JUNE 7-9, 2022 • PHILADELPHIA, PA

Small Group Discussion Topics - 2

- "Problems occur at interfaces" –A. Michael Briana
  - Consider the biggest functional and interpersonal barriers for process development and knowledge transfer, and how you have overcome them
    - Chemist/Engineer/Operating Staff,
    - R&D/Engineering/Manufacturing/Business
    - Management and stake-holders/ Technology
    - Human/machine interface (control, model, AI)
    - Vendors/customers
    - Different countries / cultures
    - You, your supervisor, and your direct reports
  - How do you quickly build trust across these interfaces for effective communication? What can be applied from our recent virtual experiences?
  - What scares your management about Process Development, and how do you deal with it?

Organized by AIChE® and the Process Development Division

PROCESS DEVELOPMENT SYMPOSIUM 2022
JUNE 7-9, 2022 • PHILADELPHIA, PA

Small Group Discussion Topics - 3

- Isn't all this "flow chemistry" stuff just sensible chemical engineering?
  - Applying the appropriate device design to best meet the process needs or is there something more?
  - What about start-up / shutdown and process upsets?
  - What about multiphase processing, especially SOLIDS…
  - Compare and contrast batch/semi-batch and continuous process development…how are the needs similar and different, and what would frighten your boss about your process choice?

Organized by AIChE® and the Process Development Division

PROCESS DEVELOPMENT SYMPOSIUM 2022
JUNE 7-9, 2022 • PHILADELPHIA, PA

Small Group 4

- SAPPHIRE gap discussion
  - critique the process development flow presented
  - what is missing?
  - which is limiting in your organization?
  - what tools do you use in each of the circles?

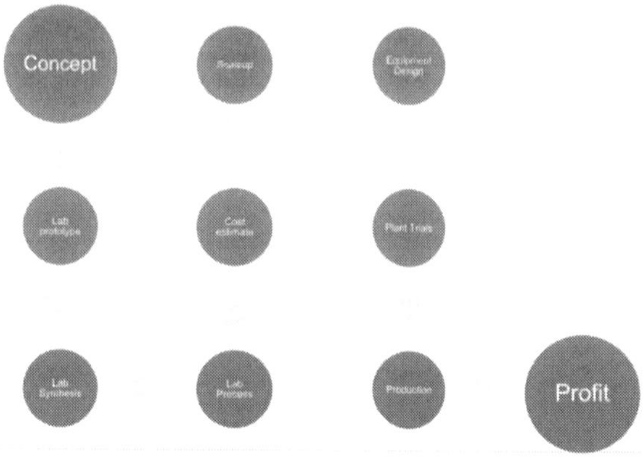

Organized by AIChE® and the Process Development Division

PROCESS DEVELOPMENT SYMPOSIUM 2022
JUNE 7-9, 2022 • PHILADELPHIA, PA

In the next 20 minutes...

- Quick demographic snapshot
- Panel Discussion
- Small Group Discussion
- Small Group Feedback

Output from this session will be included in the proceedings

Organized by AIChE® and the Process Development Division

PROCESS DEVELOPMENT SYMPOSIUM 2022
JUNE 7-9, 2022 • PHILADELPHIA, PA

Small group feedback...

- In two minutes or less
  - Who are you, your affiliation, and your discussion topic
  - What key points did your small group come up with on your discussion topic?
- Send us the details
  - You can hand us your notes (and your business card so we can ask any clarifying questions)
  - Or send us an email tonight (aaron@sarafinasprocess.com and DChakraborty2@dow.com)
  - Don't make me ask for your notes tomorrow...
- We'll summarize your notes and send them to the organizers for distribution to the attendees

Organized by AIChE® and the Process Development Division

PROCESS DEVELOPMENT SYMPOSIUM 2022
JUNE 7-9, 2022 • PHILADELPHIA, PA

Wrap up

- Enjoy the poster session and networking...

Organized by AIChE® and the Process Development Division

PROCESS DEVELOPMENT SYMPOSIUM 2022
JUNE 7-9, 2022 • PHILADELPHIA, PA

Organized by AIChE® and the Process Development Division

PROCESS DEVELOPMENT SYMPOSIUM 2022
JUNE 7-9, 2022 • PHILADELPHIA, PA

Slides to be added after session

- Acknowledgement to note takers
- Discussion topic slide 1, followed by feedback on topic
- Discussion topic slide 2, followed by feedback on topic
- Discussion topic slide 3, followed by feedback on topic
- Discussion topic slide 4, followed by feedback on topic
- Then slide the wrap up slide to the end

Organized by AIChE® and the Process Development Division

CORTEVA agriscience

Longevity studies to evaluate the performance and operating conditions of a continuous enzymatic packed bed reactor

Rick Ray, Megan Cismesia, Matt Robinson & Xiaoyong Li

June 9, 2022

Harnessing biocatalysis in process development

- The implementation of **biocatalysis** in the chemical industries have been on a drastic rise over the past several years
- **Continuous flow biocatalysis** presents numerous manufacturing advantages:
  - Reduced environmental impact
  - Improved heat & mass transfer
  - High reaction selectivity

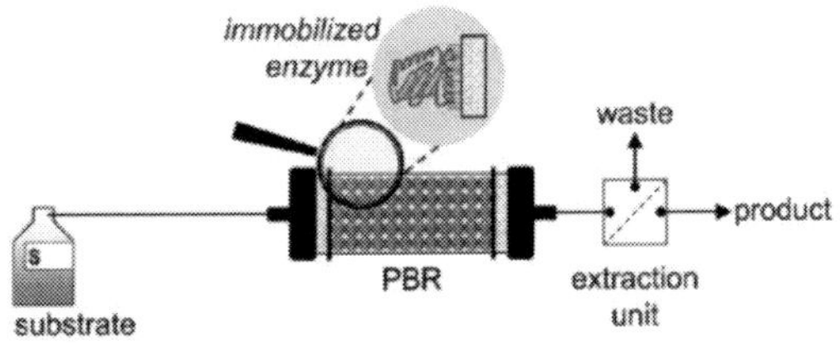

Example of a continuous flow biocatalysis system using a packed bed reactor with an immobilized enzyme*

 *De Santis P., Meyer L-E & Kara S. (2020) *React. Chem. Eng.*, 5, 2155-2184

Enantioselective enzymatic resolution

- **Enzymatic** kinetic resolution involves the use of a biocatalyst to discriminate between enantiomers in a racemic mixture
- Process discriminates undesired enantiomer by acetylation mechanism using immobilized lipase beads
- Initial processing was performed in a stirred tank reactor resulting in significant bead swelling and fracturing

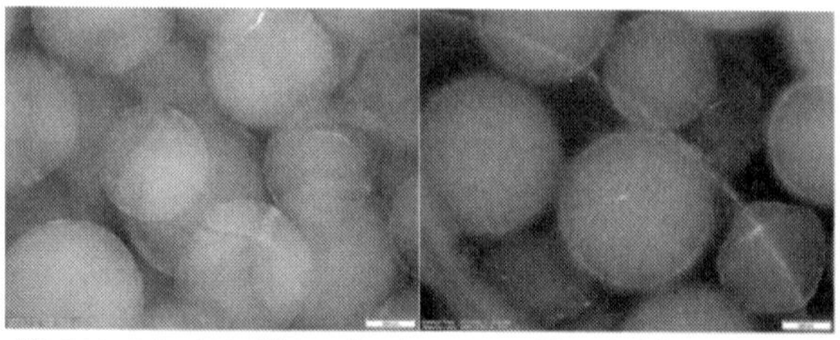

Microscope images of lipase beads recovered from batch reactor system

General block flow diagram for enzymatic resolution in stirred tank system

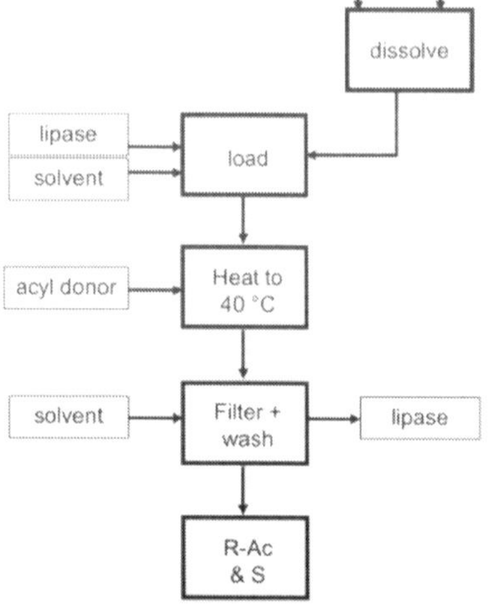

Enantioselective enzymatic resolution

- **Enzymatic** kinetic resolution involves the use of a biocatalyst to discriminate between enantiomers in a racemic mixture
- Process discriminates undesired enantiomer by acetylation mechanism using immobilized lipase beads
- Initial processing was performed in a stirred tank reactor resulting in significant bead swelling and fracturing

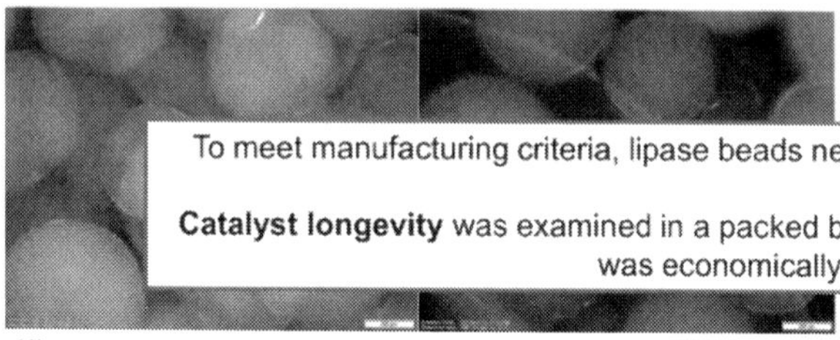

Microscope images of lipase beads recovered from batch reactor system

General block flow diagram for enzymatic resolution in stirred tank system

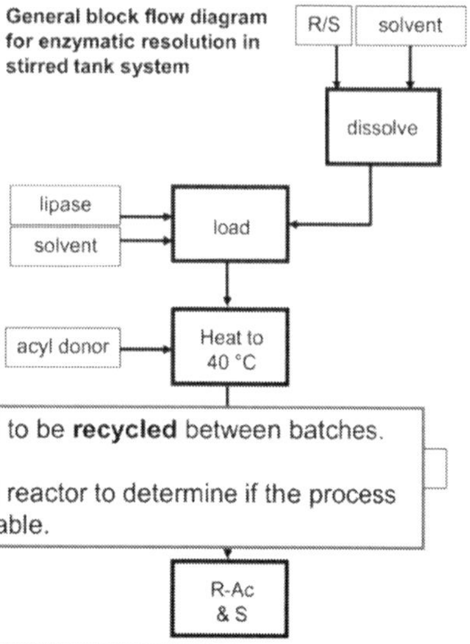

To meet manufacturing criteria, lipase beads need to be **recycled** between batches.

Catalyst longevity was examined in a packed bed reactor to determine if the process was economically viable.

Project overview

1. REACTOR ASSEMBLY

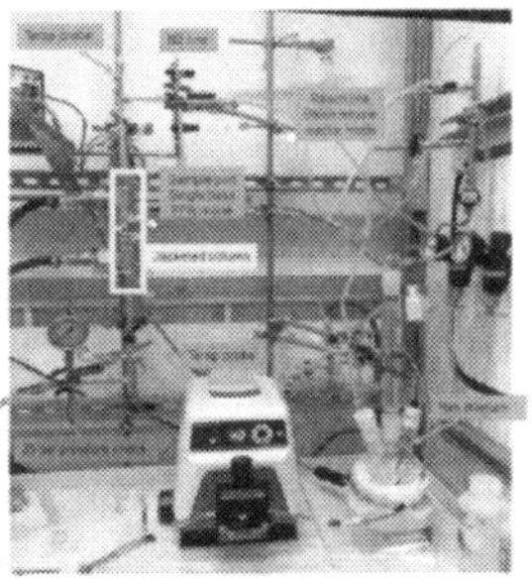

2. COLUMN PACKING

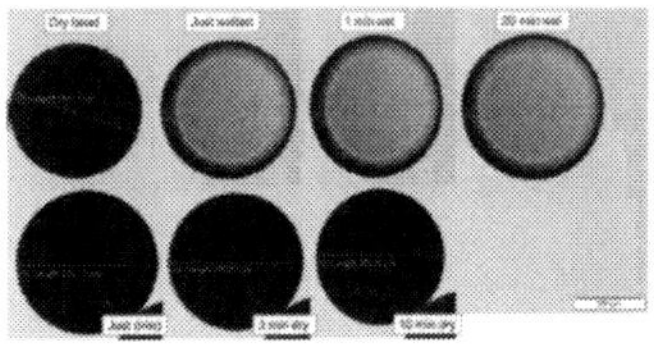

3. OPERATING CONDITIONS

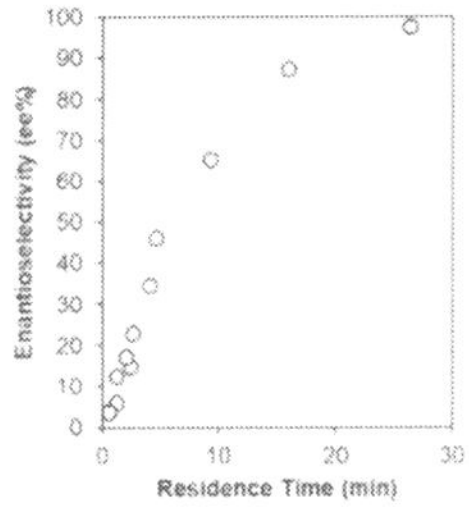

4. ENZYME LONGEVITY STUDIES

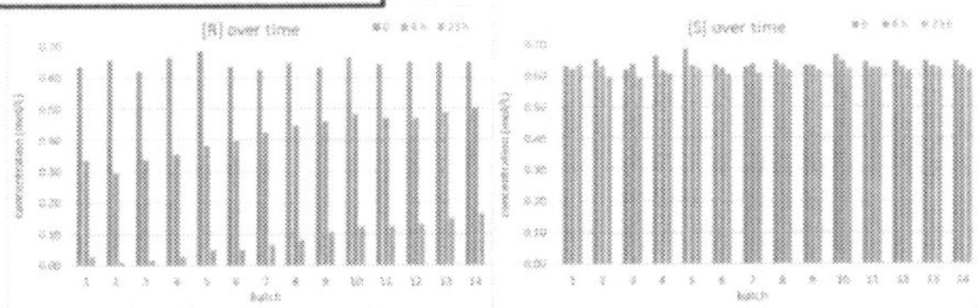

Packed bed reactor assembly

- Packed bed column was a ½" stainless steel tube with ~21 mL of empty volume to be packed with lipase beads
- This tube was enclosed by 1" Swagelok and stainless-steel tubing to allow for suitable heat transfer control around the packed bed for isothermal operation

Packed column design.
Left: Profile view of packed column
Right: View of top & bottom of column packed with glass wool

Flow operation

- The packed column was integrated as part of a flow system controlled by a peristaltic pump
- A three-necked round bottom flask served as the holdup for the reaction mixture
- Heating tape was used to ensure mixture remains at a consistent temperature
- The system could be operated in two modes:
  - Single pass/plug flow (PFR) mode
  - Batch recycle reactor mode

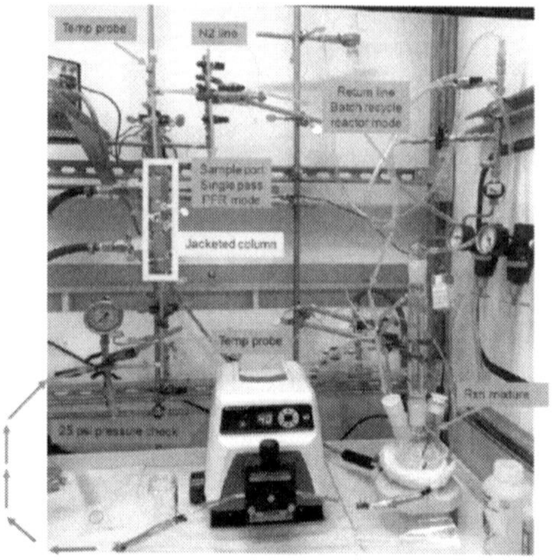

Flow reactor system schematic
(Red arrows indicate direction of flow)

CORTEVA agriscience

7

Column packing considerations

- Column overpacking leads to fracturing of enzyme beads and **loss in activity**
- Beads experience substantial swelling in the choice solvents. Findings were consistent with supplier specifications which indicate **~50% bead swelling over 24 h of submersion**
- Accounting for bead swelling was a necessary step in determining the volume of dried beads to be packed in the column

CORTEVA agriscience

8

Microscopy of dry beads

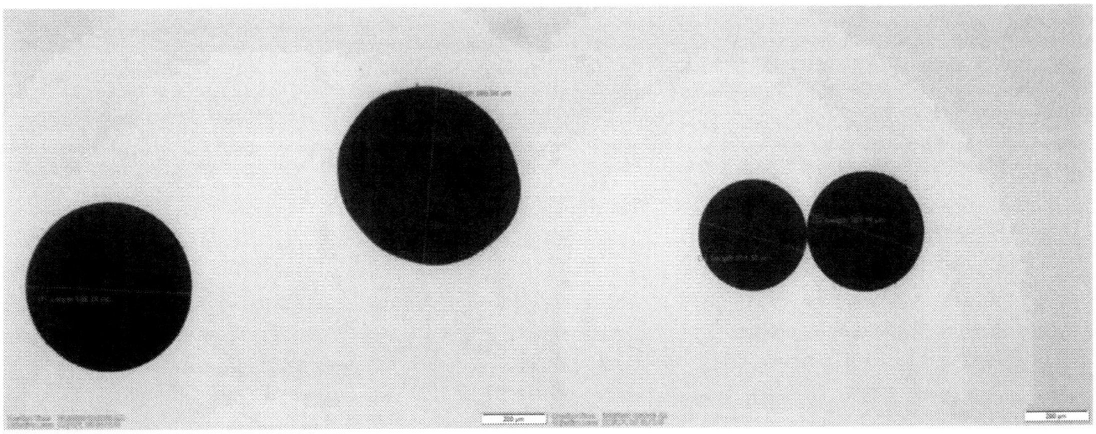

CORTEVA agriscience 9

Microscopy of wetted beads

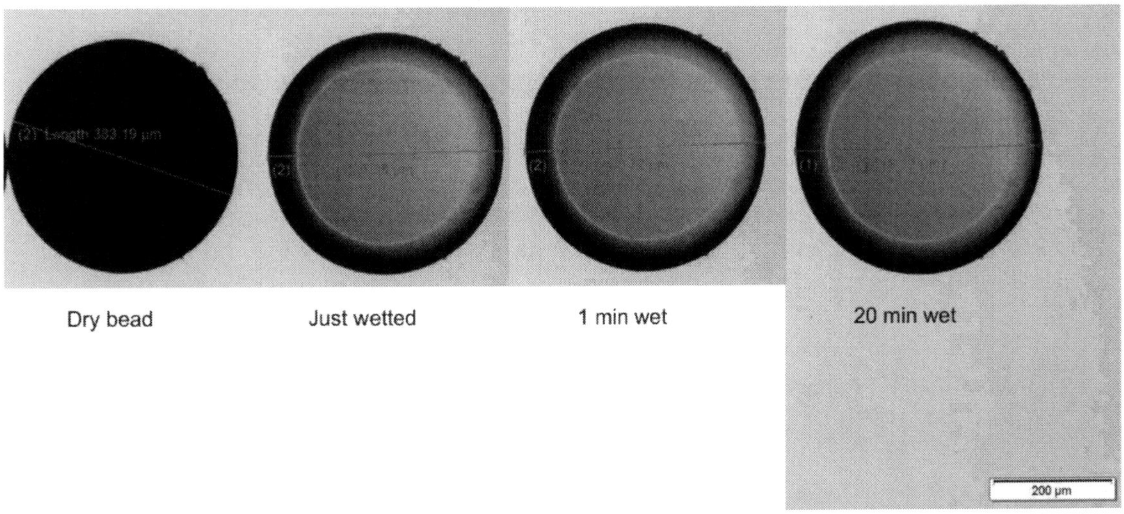

Dry bead | Just wetted | 1 min wet | 20 min wet

CORTEVA agriscience 10

Microscopy of wetted and redried beads

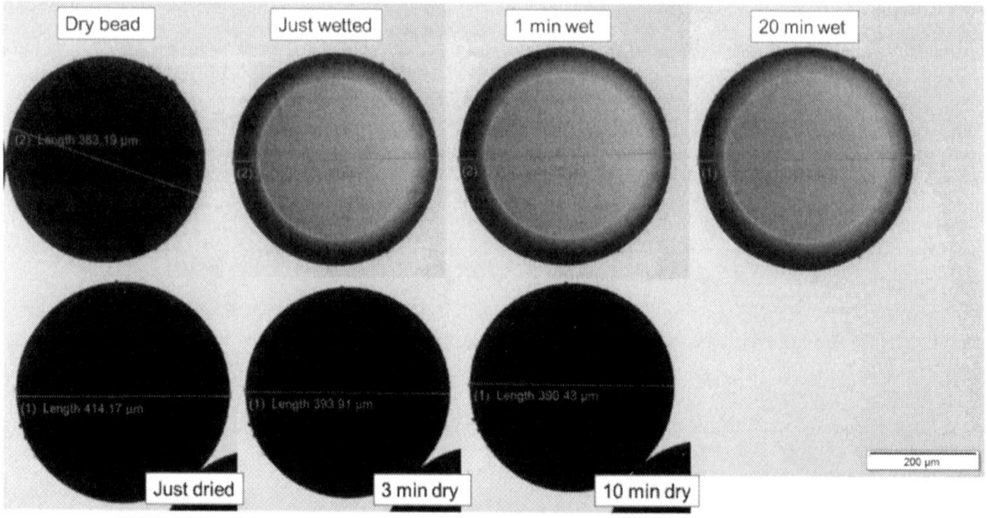

Bead swelling in reaction mixture

- ~56% height difference after 20 min of wetting
- Extent of swelling plateaus after 20 mins
- Dry bead density was determined to be ~0.27 g/mL, and thus we could estimate the mass of dry beads necessary to pack the column

Flow rate selection

System was operated in single pass to obtain ideal flow rate for longevity studies

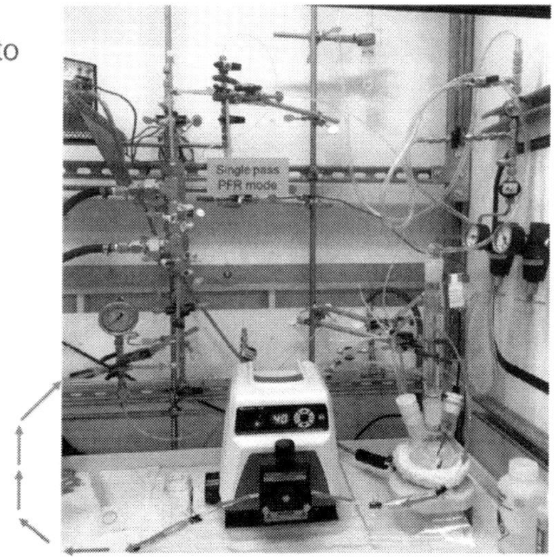

CORTEVA agriscience

13

Flow rate selection

Data from single pass operation under varying flow rates

Pump speed (rpm)	Flow rate (mL·min^{-1})	Residence time (min)	Enantiomeric excess (ee%)	Reynolds number[a]	Reynolds number[b]
3	0.3	26.3	97.6	0.02	0.06
5	0.5	16	87.5	0.04	0.1
10	0.8	9.3	65.4	0.07	0.17
20	1.7	4.6	46.1	0.14	0.35
20	1.9	4	34.6	0.16	0.4
40	3.1	2.4	15	0.26	0.67
40	2.9	2.6	22.9	0.25	0.62
40	3.8	2	17.2	0.35	0.81
80	6.1	1.2	12.6	0.52	1.31
80	6.3	1.2	6.2	0.54	1.36
160	12.7	0.6	3.8	1.1	2.74
160	13.3	0.6	3.9	1.15	2.85
160	12.8	0.6	3.7	1.1	2.75

[a]Reynolds number as defined in Rhodes M. (1989) *Introduction to Particle Technology*. John Wiley & Sons

$$\mathrm{Re} = \frac{\rho v_s D}{\mu}$$

[b]Reynolds number as defined in Dwivedi PM and Upadhyay SN. (1977). *Ind. Eng. Chem. Process Des. Dev.*, 16, 157-165

$$\mathrm{Re} = \frac{\rho v_s D}{\mu \varepsilon}$$

Where ρ is the fluid density, v_s is the superficial velocity, D is the particle diameter, μ is the fluid dynamic viscosity, and ε is the void fraction

- System was operated in single pass at varying flow rates to determine effect of residence time on enantioselectivity
- As expected, high conversion was observed under longer residence times
- All flow rates tested resulted in laminar flow operation

CORTEVA agriscience

14

Flow rate selection

Data from single pass operation under varying flow rates

Pump speed (rpm)	Flow rate (mL·min^{-1})	Residence time (min)	Enantiomeric excess (ee%)	Reynolds number[a]	Reynolds number[b]
3	0.3	26.3	97.6	0.02	0.06
5	0.5	16	87.5	0.04	0.1
10	0.8	9.3	65.4	0.07	0.17
20	1.7	4.6	46.1	0.14	0.35
20	1.9	4	34.6	0.16	0.4
40	3.1	2.4	15	0.26	0.67
40	2.9	2.6	22.9	0.25	0.62
40	3.8	2	17.2	0.35	0.81
80	**6.1**	**1.2**	**12.6**	**0.52**	**1.31**
80	**6.3**	**1.2**	**6.2**	**0.54**	**1.36**
160	12.7	0.6	3.8	1.1	2.74
160	13.3	0.6	3.9	1.15	2.85
160	12.8	0.6	3.7	1.1	2.75

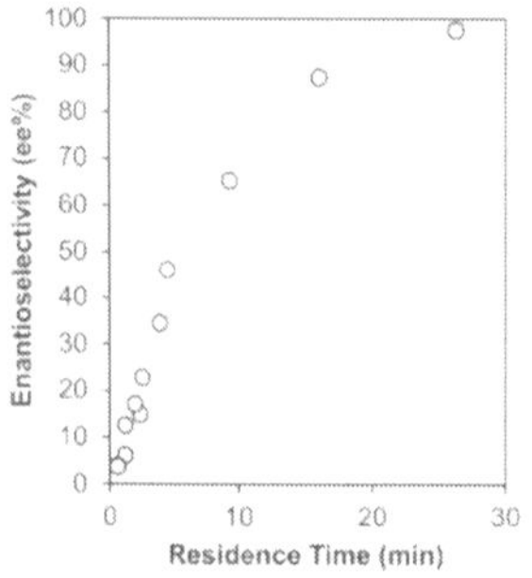

- Clear mass transfer limitations observed for longer residence times
- Batch recycle mode should improve throughput using a higher flow rate
- A flow rate of ~6 ml/min was selected to use for the batch recycle mode.

CORTEVA agriscience

15

Longevity studies

Catalysis longevity studies were performed in batch recycle mode (14 batches, 75 g scale) with 6.7 wt% bead loading to reach desired throughput and conversion according to manufacturing considerations

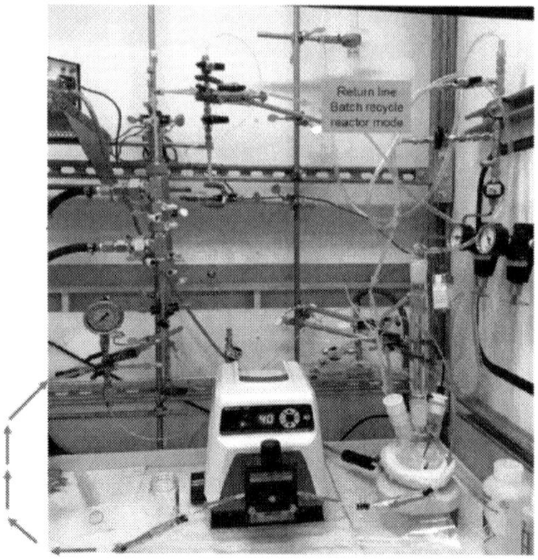

CORTEVA agriscience

16

Longevity study results

- Loss in productivity was measured across 14 batches. Each batch was carried out over a period of one day.
- Between each batch, the column was purged with solvent. When not in use, the column was left wetted with solvent.
- While desired conversion is met across all batches, there was a clear decrease in enantioselectivity with successive batches.

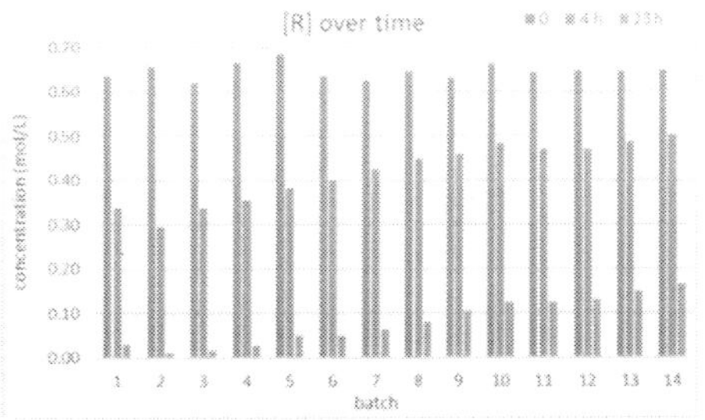

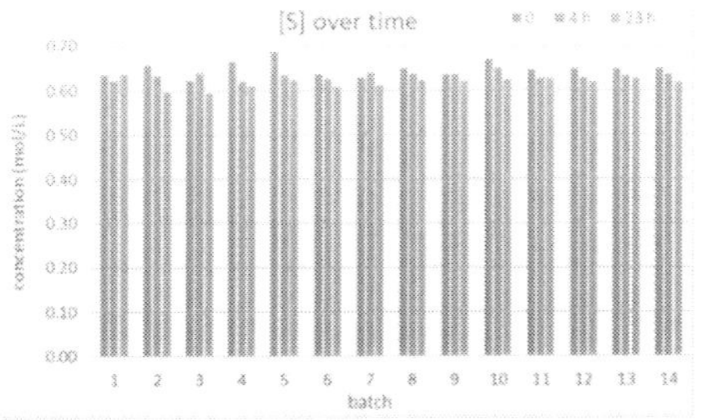

Concentrations of undesired enantiomer (left) and desired product (right) at three time points: 0, 4 and 23 h of operation

Determining total conversion time

- Due to loss in activity across successive batches, one batch (Batch 11) was allowed to run for longer (53 h) to achieve full chiral resolution
- Using measured concentrations vs. time, the 'time to completion' was estimated based on first order kinetics and assuming 98% end-point conversion:

$$\textbf{Time to completion} = \mathrm{m} \cdot [\ln(0.02 \cdot R_{t=0}) - b]$$

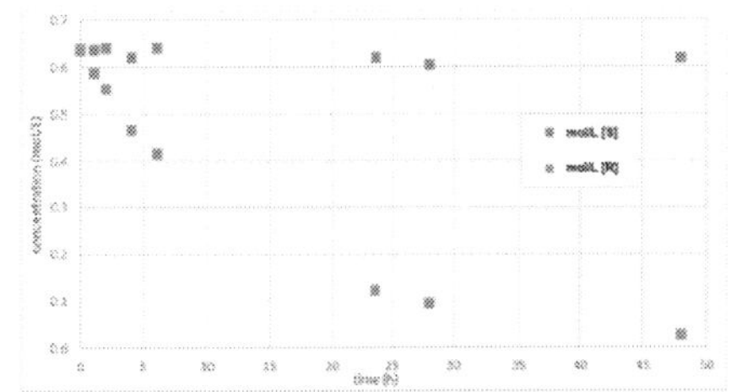

Concentration profiles for the enantiomers for Batch 11

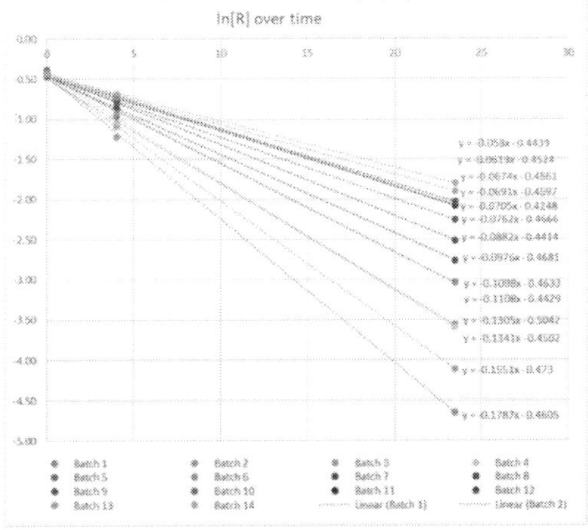

Natural logarithmic concentration profiles for undesired enantiomer to estimate 'time to completion'

Summary and conclusions

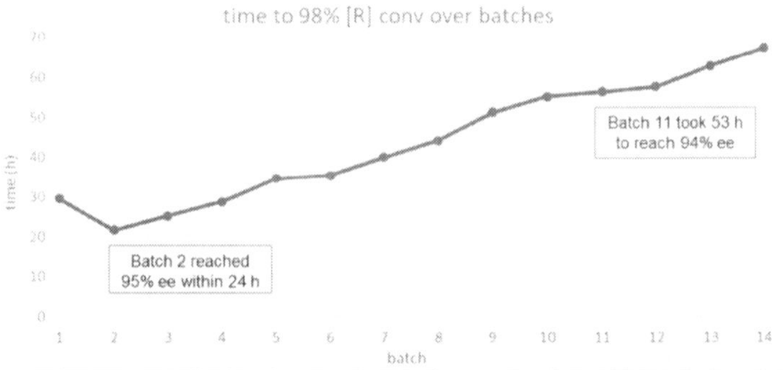

Calculated time to 98% conversion of [R] based on a good fit of apparent first order kinetics
Reactions were only run to 23.5 h except for batch 11. [R] is concentration of the undesired enantiomer of the starting material.

- The lipase beads demonstrate sufficient activity after 14 batches, but productivity steadily decreases after the second batch.
- No plateau in productivity loss observed, suggesting that the steady loss in enantioselectivity will continue to decrease with subsequent batches.
- Beads can be recycled for requisite number of batches to meet desired selectivity specifications, but it requires adjustments in either flow rate or total processing time to meet throughput demands.
- This observed degradation suggests additional research necessary to minimize biocatalyst degradation rate.

Exometabolome profiling reveals the activation of the carnitine buffering pathway in fed-batch cultures of CHO cells co-fed with glucose and lactic acid

Keegan Orzechowski and Juan Aon

GlaxoSmithKline

R&D BioPharma Process Development

King of Prussia, PA

June 9th, 2022

Outline

- BACKGROUND:
  - QbD paradigm
  - How 'Omics approaches interact – Central Dogma of Biology
- INTRODUCTION:
  - Current knowledge
- EXPERIMENTAL DESIGN:
  - Cellzome platform, data quality control, and PCA analysis
- RESULTS:
  - Ammonia metabolism via transamination reactions
  - Acetyl group removal via Carbon-buffering system mediated by carnitine
  - Alternative removal by HMG and Mevalonate pathways
  - TCA cycle functioning
- CONCLUSIONS
- Q & A

Background

Relevant studies…

This initiative started with the study of the impact of CHO cell's lactate assimilation on the glycosylation pattern of a mAb:

- *Tuning Monoclonal Antibody Galactosylation using Raman Spectroscopy-Controlled Lactic Acid Feeding. Eyster, Talwar, Fernandez, Foster, Hayes, Allen, Reidinger, Wan, Ji, Aon, Patel, Ritz, Biotechnology Progress, 2020*

Characterization of CHO cell's metabolism and its impact:

- Enhance our knowledge of the current platform cell line CHO-K1
  - Deep dive in the CHO cell physiological adjustment/adaptation with lactic acid feed
  - Understanding biomarkers and surrogate metabolites of activated pathways
  - Meaning of activated metabolic pathways for CHO cell homoeostasis to grow under fed-batch conditions

The characterization approach based on metabolome profiling:

- High-throughput screening of metabolites
- Un-biased/un-targeted profiling
- In-house experience for data analysis and interpretation

Know-how of the CHO cell micro-factory provides a better understanding on the production process performance and intended product quality profile

Quality-by-Design (QbD) paradigm:

- ❑ This approach aims to: "Once the significant parameters are identified, they can be further studied (e.g., through a combination of design of experiments, mathematical models, or studies that lead to mechanistic understanding) to achieve a higher level of process understanding".

*2009 *ICH* Q8(R2) Pharmaceutical Development

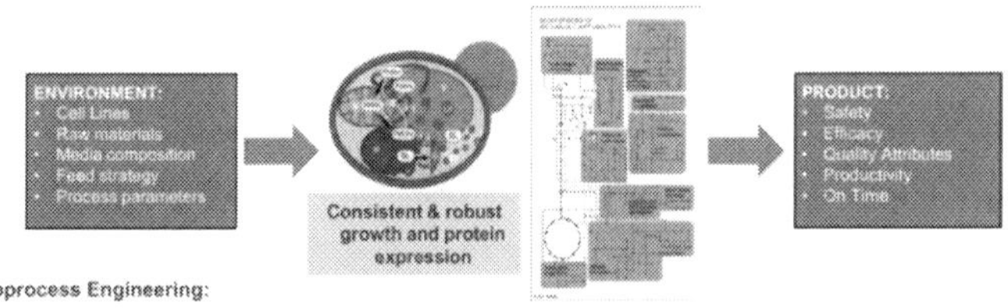

❑ **Bioprocess Engineering:**

- ➢ Understand the interaction shown as a particular phenotype between the organism and its environment
- ➢ How that interaction impacts process performance, its consistency, and product quality
- ➢ Know-how for platform understanding and baseline for future molecules to be expressed

How 'Omics approaches interact – Central Dogma of Biology

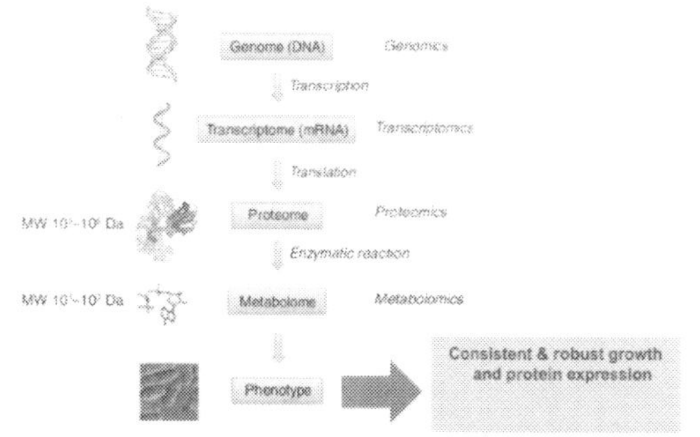

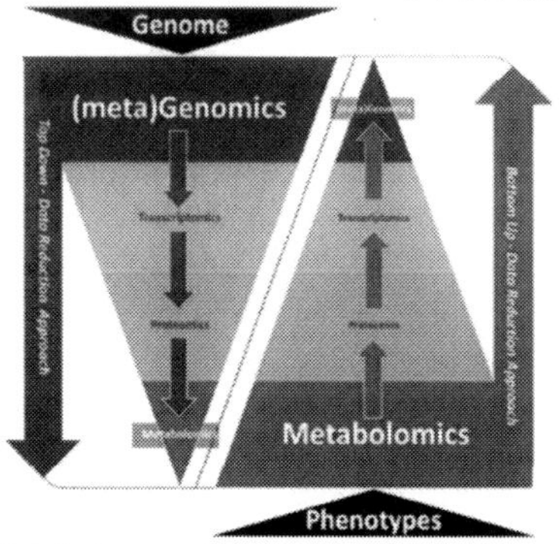

Pluskal, Tomáš & Yanagida, Mitsuhiro. (2016). Metabolomic Analysis of Schizosaccharomyces pombe : Sample Preparation, Detection, and Data Interpretation. Cold Spring Harbor Protocols. 2016. pdb.top079921. 10.1101/pdb.top079921.

Pinu, F.R., Beale, D.J., Paten, A.M., Kouremenos, K., Swarup, S.; Schirra, H.J.; Wishart, D. Systems Biology and Multi-Omics Integration: Viewpoints from the Metabolomics Research Community. Metabolites 2019, 9, 76. https://doi.org/10.3390/metabo9040076

Background

Scenario of the problem, preliminary results, and working hypothesis

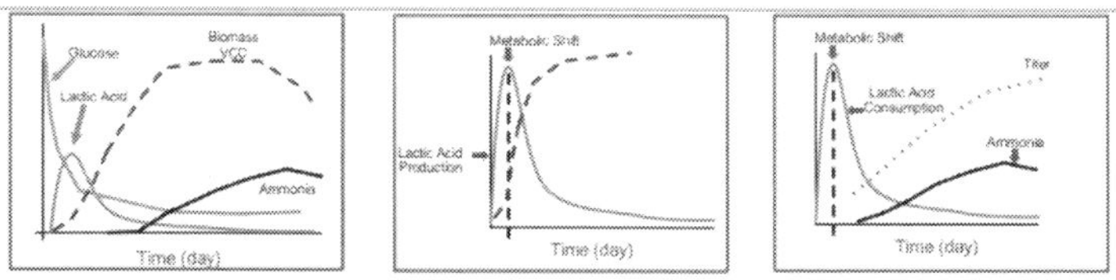

- Characterization of the metabolic adjustment after the "*Lactate-shift metabolism*":
  - Alternative pathways fueling the TCA (Krebs) cycle such as amino acid catabolism and fatty acid β-oxidation
  - Novel finding of activation of the carbon-buffering pathways involving carnitine, HMG, and mevalonate
- Working hypothesis: The oxidative phosphorylation metabolism is required after "*Lactate-shift metabolism*" and allows CHO cells to have better ATP/redox yield and could sustain a better growth and/or productivity.

Exponential phase day, 0 to 5

Aerobic glycolysis:

-Exclusive quick consumption of glucose and secretion of by-product lactate.

-Over-flow metabolism known as Warburg phenomenon.

-Glycolysis until lactate production generates 2 ATPs per glucose molecule consumed, redox neutral

Stationary and decline phases, day 5 to 17

Preponderant Oxidative Phosphorylation (OXPHOS) metabolism:

-Co-assimilation of glucose and lactate.

-Additional supply of intracellular pyruvate from lactate.

-~36 ATPs per glucose molecule fully catabolized ("burned") via OXPHOS.

-Our study: Metabolic adaption based on several biomarkers found by a un-biased/un-targeted Exometabolome profiling

Source: Templeton et al., 2013, 2017; Hartley et al., 2017

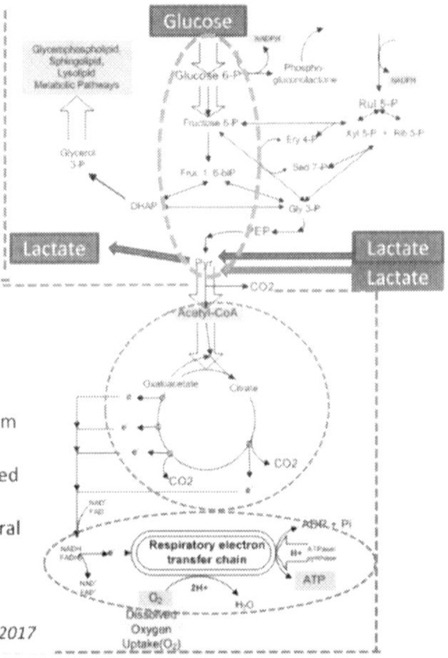

Feedback Control Loop for Continuous Feeding

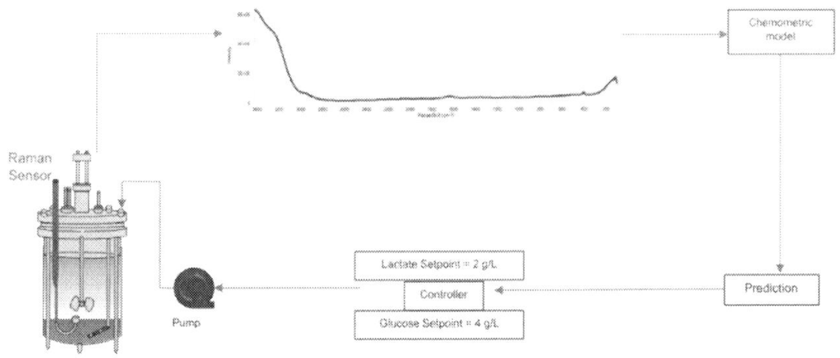

Levels of lactic acid detected confirms treatment group assignment

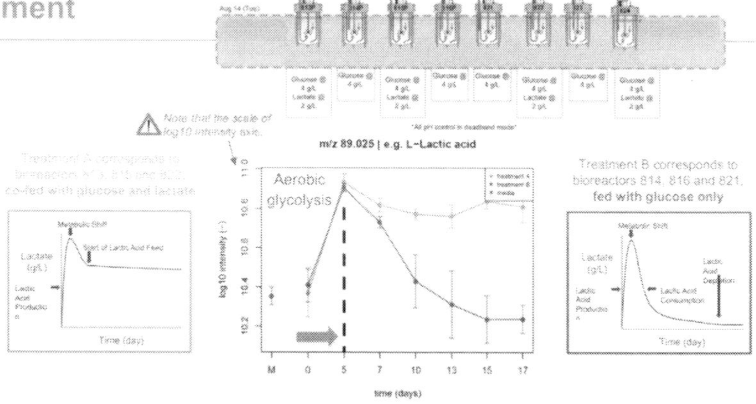

The lactate levels detected in both treatment groups over 17 production days (day 0, 5, 7, 10, 13, 15, and 17) correspond to expected lactate concentrations in both group and allowed to confirm of group assignment.

9

Metabolome profiling by Cellzome

High-throughput discovery metabolomics at Cellzome

Platform based on Fuhrer et al. (2011), Anal.Chem., doi:10.1021/ac201267k

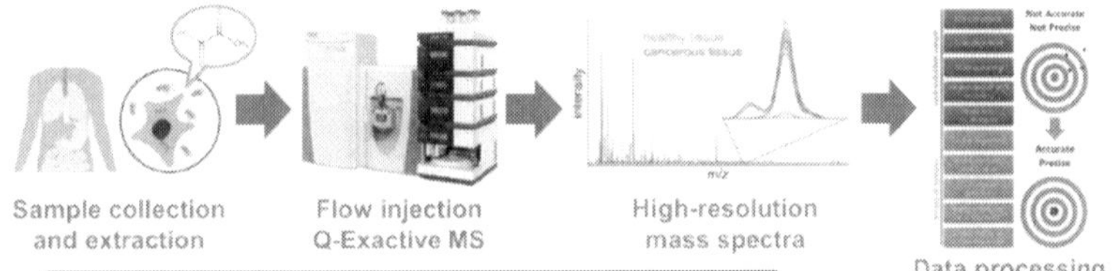

Key platform features:

- **High throughput** – 1,000 samples/day
- **Semi-quantitative** – fold-changes relative to reference
- **Broad coverage** – Typically ~10,000 ions / ~3,000 polar metabolites
- **Tentative metabolite ID** – Accurate mass and isotopologues

Sample requirements (compatible with diverse matrices):

- **Tissues:** ≥10 mg
- **Body fluids:** ≥10 µL
- **Cells:** ≥1E6 → potential to scale down to ~1E4 cells/sample
- **Media:** ≥5 µL

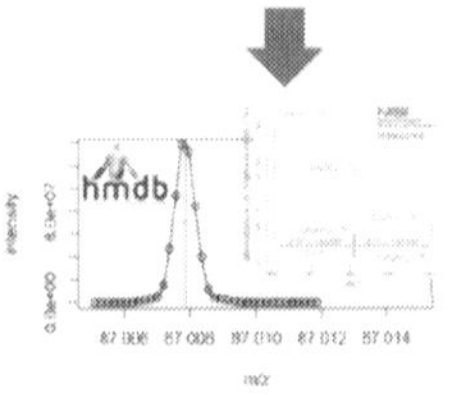

General overview of samples shows different metabolic profiles between A and B at the highest time points

Principal Component Analysis (PCA) based on overall metabolic profiles of all samples:

- Ions filtered for log10 intensity ≥ 2
- Each number on plot represents one sample (number = corresponding production day)

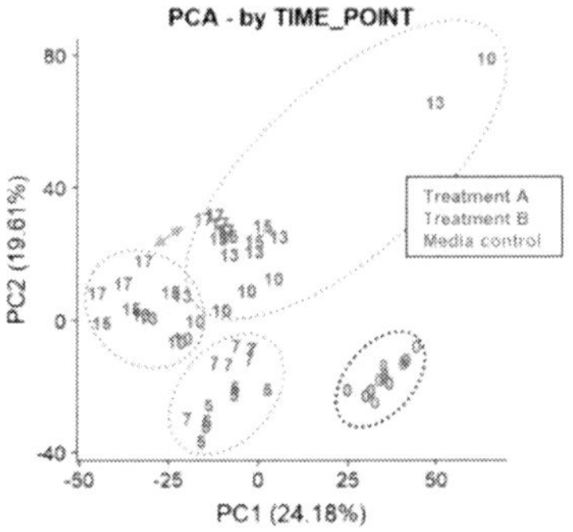

- All day 0 samples cluster together:
  - Similar metabolic profiles between cell supernatants and media control at day 0.
- Treatments A and B cluster together at the lower time points (day 5 and 7):
  - Similar metabolic profiles until day 7.
- Treatments A and B cluster apart from each other at the higher time points (day 10 to 17):
  - Difference in overall metabolic profiles between A and B can be observed at day 10 and above.
- 2 samples seem to cluster outside of the main population (day 10 and 13)
  - These 2 outliers were kept for the statistical analysis and don't seem to influence the results significantly.

Source: Vappiani, Eyster, Orzechowski, Ritz, Patel, Sevin, Aon, Biotechnol. Progress 2021

CHO exometabolome results:

1- Growth and viability profiles

2- mAb titer profiles

3- Ammonia secretion profiles

4- Ammonia metabolism via transamination reactions

Cell physiological responses: VCC and Viability

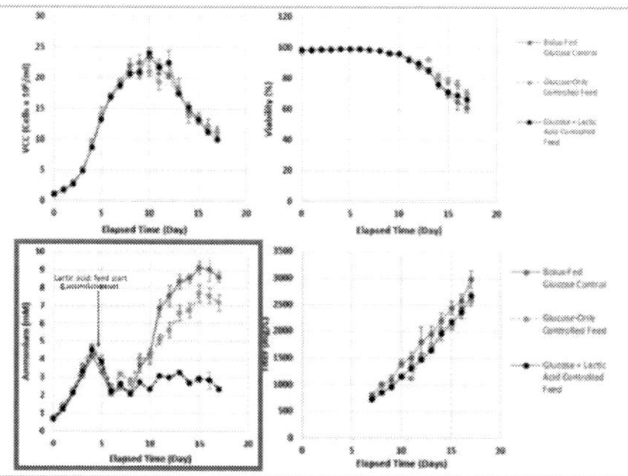

Source: Eyster *et al.* 2020; Biotechnol. Progress

Ammonia metabolism via transamination reactions

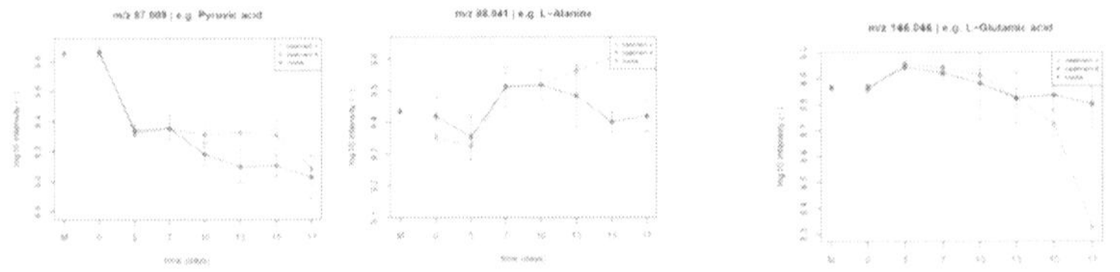

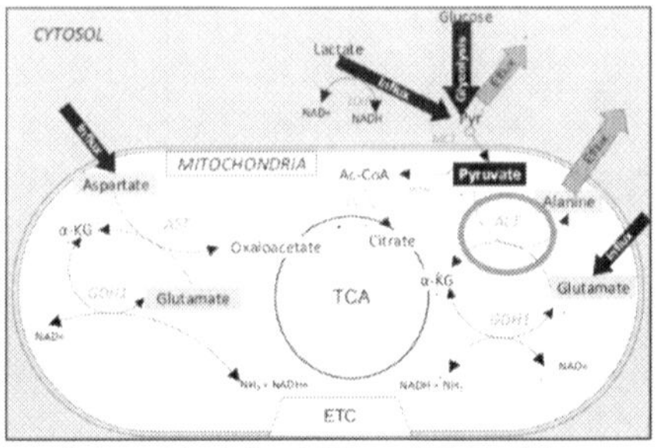

Reference: Li *et al.*, Biotech & Bioeng (2012) 109: 1173

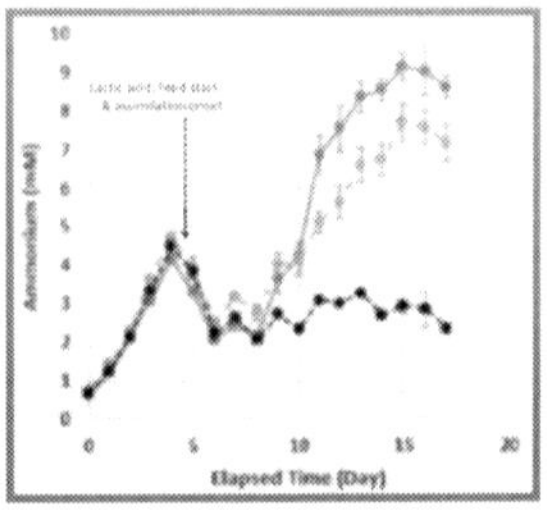

Source: Eyster *et al.* 2020, Biotechnol. Progress

Novel results from CHO exometabolome:

1- Main metabolic pathways

2- Acyl-group removal via Carbon-buffering system mediated by carnitine

3- Alternative removal via HMG and Mevalonate pathways

4- TCA cycle functioning

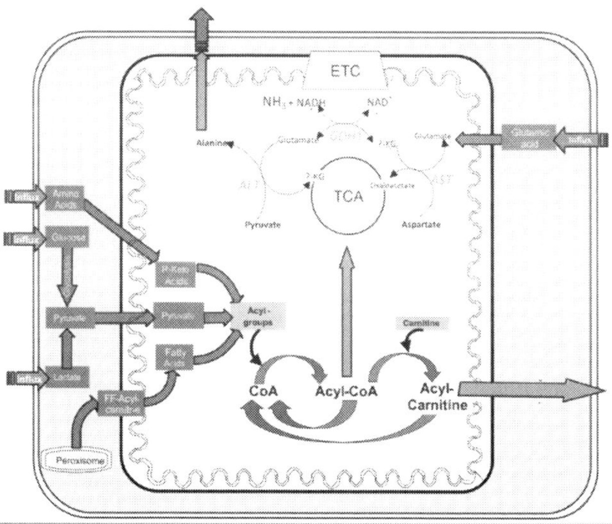

References: Davies M. et al., 2016, Cell Reports 14, 243-254; Li et al., Biotech & Bioeng (2012) 109: 1173

Carbon-buffering system based on carnitine: key biomarkers of incomplete Fatty Acid β-oxidation

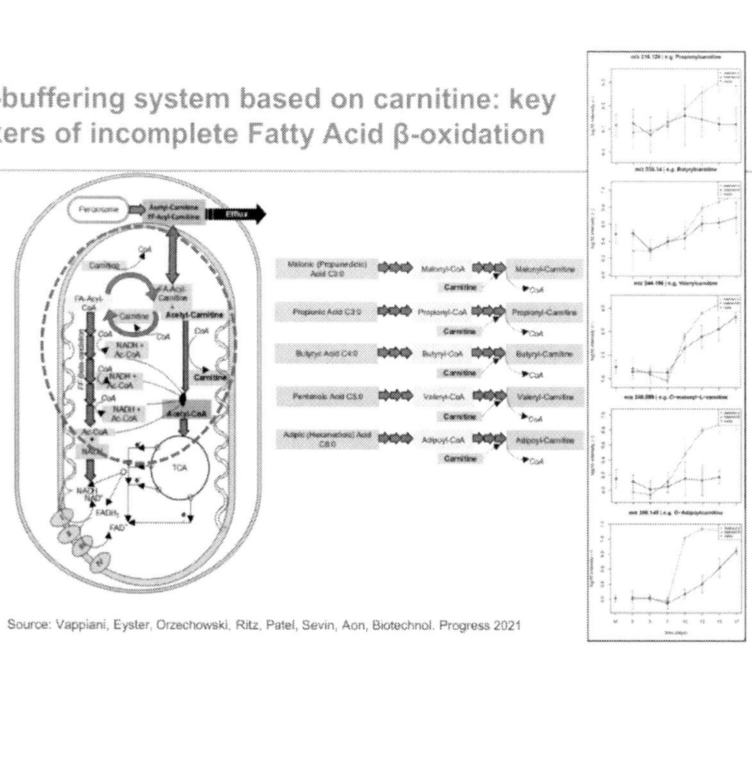

Source: Vappiani, Eyster, Orzechowski, Ritz, Patel, Sevin, Aon, Biotechnol. Progress 2021

Carbon-buffering system based on carnitine: key biomarkers of amino acid anaplerotic catabolism

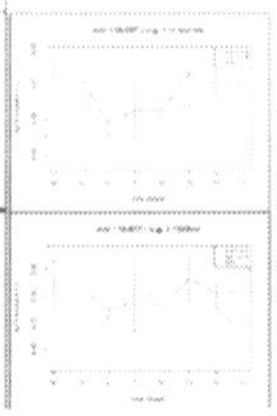

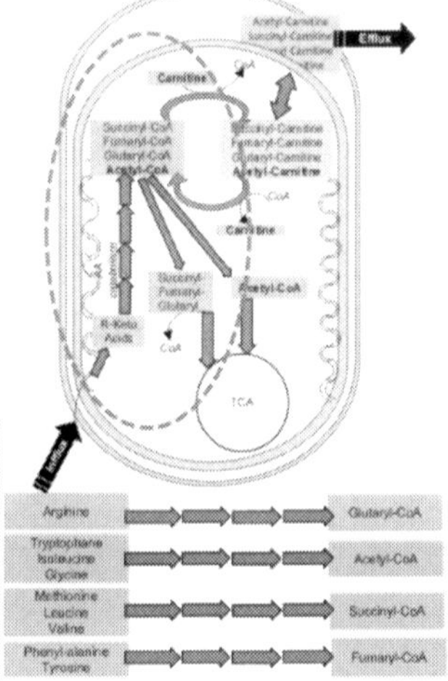

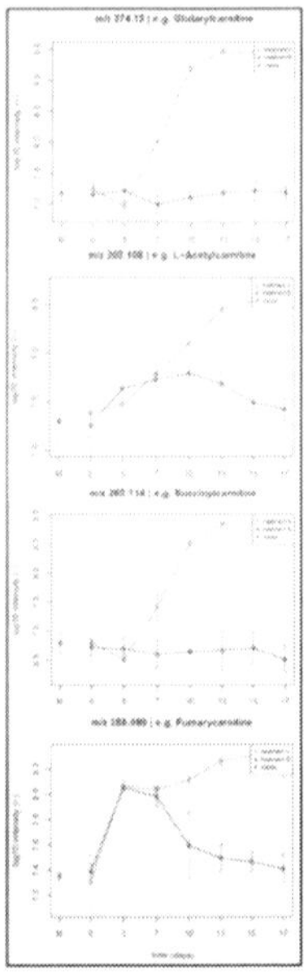

Source: Vappiani et al. Biotechnol. Progress 2021

Alternative acetyl group removal pathways: HMG and Mevalonate

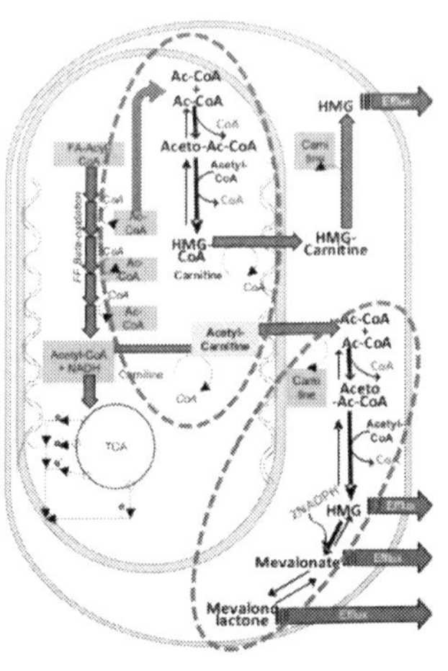

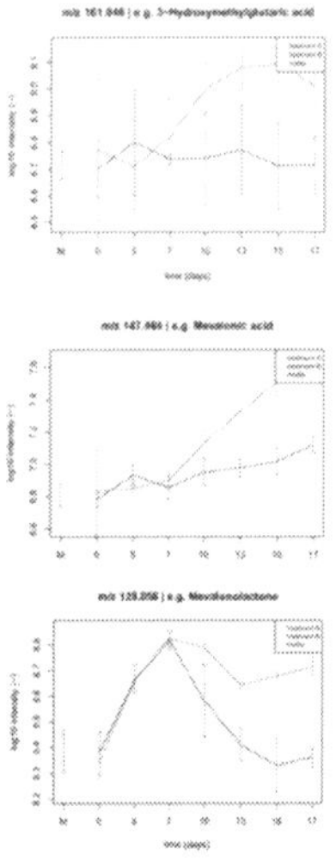

Carnitine-dependent transport. Roermund *et al.*, 1999 The EMBO Journal Vol. 18:5843-5852

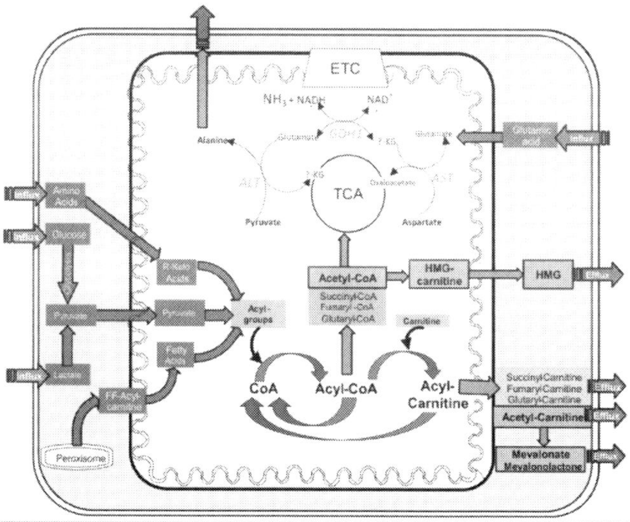

Source: Vappiani, Eyster, Orzechowski, Ritz, Patel, Sevin, Aon, Biotechnol. Progress 2021

TCA cycle functioning comparable between treatments A and B

Metabolic shift to lactic acid consumption:

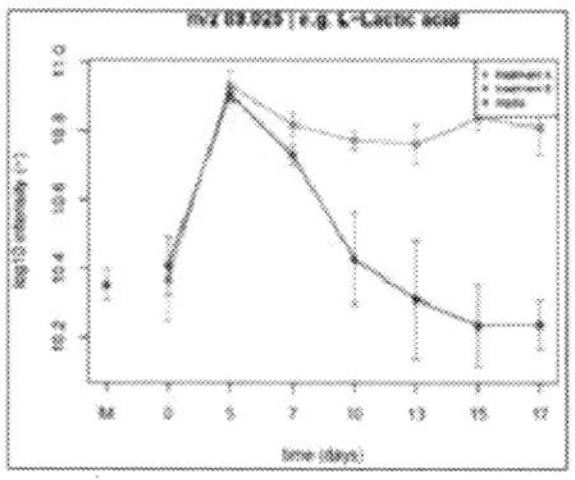

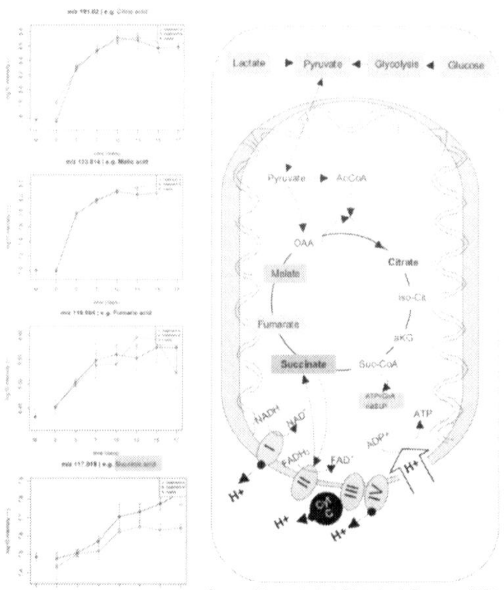

Source: Vappiani et al. Biotechnol. Progresa 2021

Conclusions

- There is evidence of the following metabolic adjustments when Lactate was fed:
  - We found the simultaneous activation of anaplerotic pathways of the TCA cycle as well as of the carbon-buffering pathway mediated by carnitine.
  - CHO cells exhibited a resulting balance between the supply of intermediates from (*i*) amino acid oxidative catabolism, (*ii*) the β-oxidation of fatty acids, and (*iii*) pyruvate from glycolysis and lactic acid; and the secretion of their intermediate-carnitine derivatives from the carbon-buffering system to keep the TCA cycle properly functioning.
  - Together, the data show that carnitine buffering, HMG, and mevalonate pathways are part of a known mechanism in mammalian cells to overcome a limited TCA cycle capacity to assimilate those acyl-CoA groups or to maintain the acyl-CoA/free CoA ratio for proper and continuous mitochondrial function.

  - **Final comment: the proper function of the mitochondria, mainly the TCA cycle and Electron Transfer Chain (ETC), are key highly demanded pathways after the metabolic shift which allow CHO cells to grow and produce mAbs.**

Acknowledgements

Drug Substance Process Development:
- Tom Eyster
- Diana Ritz
- Shelby Foster
- Randy Allen
- Scot Reidinger
- James Hayes
- Jan Fernandez
- Sybille Galosy
- Prem Patel

Metabolomics:
- Johanna Vappiani
- Daniel Sevin

Opportunities and Challenges in the Biocatalytic Process Development of APIs

Rakeshwar Bandichhor, PhD
Vice President and Head of Chemistry, API-PR&D

7 to 9th of June 2022

Dr.Reddy's

Caveats

The views expressed are personal and do not necessarily reflect those of my employer or any other organisation with which I am affiliated.

A Vibrant Trillion Dollar Plus Health Industry (Mainly Pharma); Growing steady globally

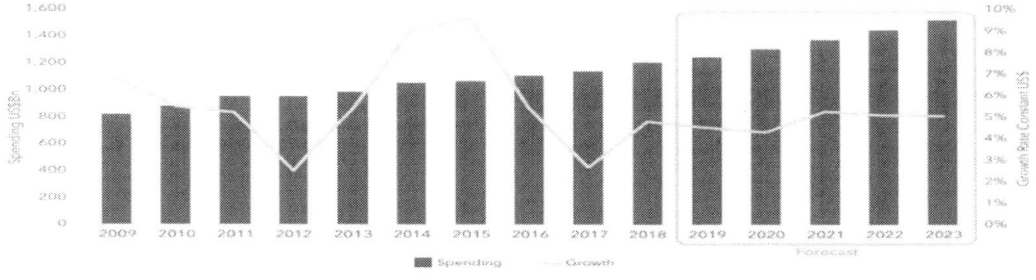

Source: IQVIA Market Prognosis, Sep 2018; IQVIA Institute, Dec 2018

- As of 2018, approximately 1.2 trillion USD had been spent on medicines, up from just 887 billion U.S. dollars in 2010. That number is expected to increase to 1.52 trillion USD by the year 2023
- Global revenue in 2018 was at 1.2 trillion USD where as in 2001 the market was valued at 390 bn USD
- The industry is expected to grow by 160% between 2017 and 2030

India: Pharmacy to the World

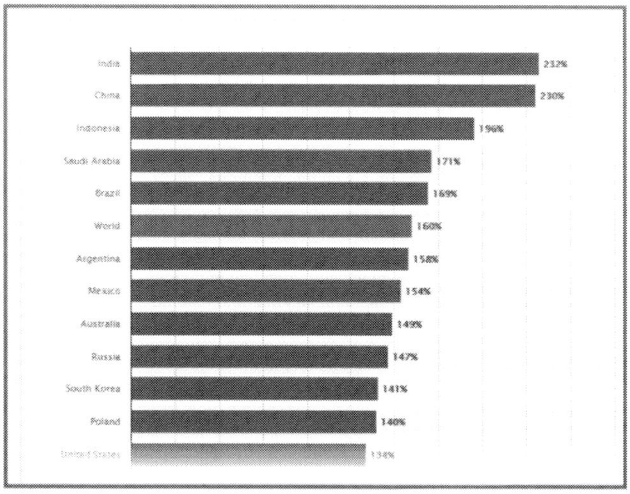

Worldwide forecast of pharmaceutical sector growth between 2017 and 2030 by country

Source: www.statista.com/statistics/783145/pharmaceutical-sector-growth-forecast-worldwide-by-country/

- Indian Pharma market stood at USD 30 Billion in 2018 and is expected to grow to USD 55 billion by 2020 (McKinsey, Indian Pharma 2020 Report)
- Globally, the Indian Pharma market is the third largest in terms of volume and 13th in terms of value (The Hindu, Business Line)
- Every fifth generic drug in the world comes from India; Every third prescription in US is filled by a made-in-India drug
- Indian firms have over 40% of total generic drug approvals in US
- 70-90% of essential medicines of developing countries come from India

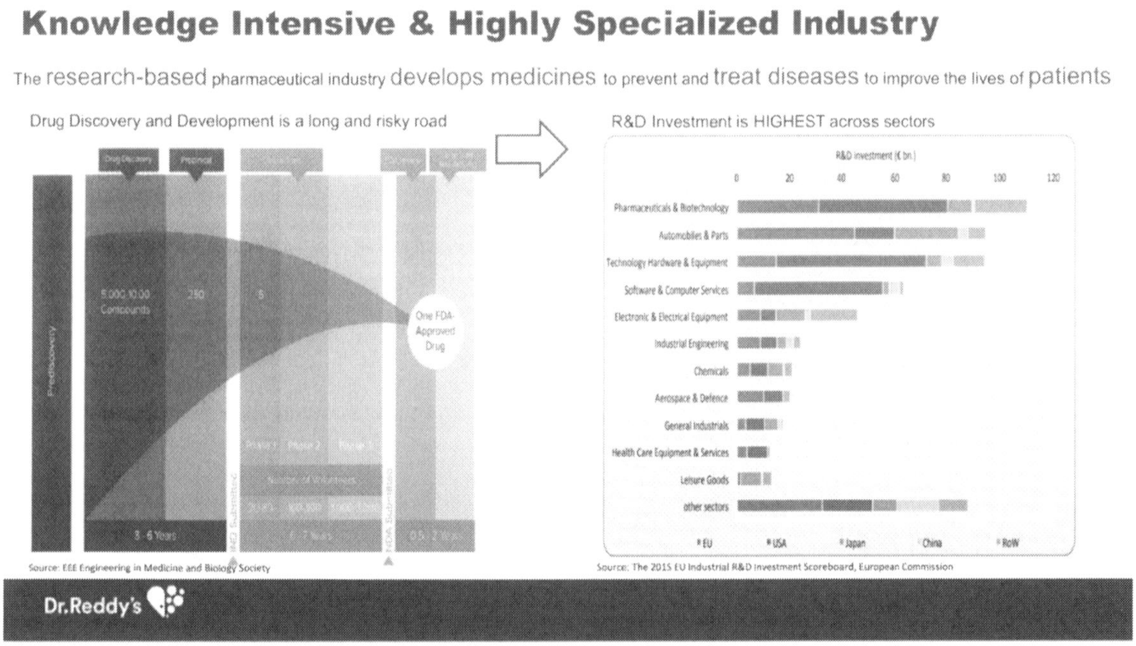

Opportunities

"In order to have faster and broader implementation of enzymatic steps in API syntheses, several opportunities are to be leveraged"

1. For biocatalytic processes, it must be considered that water is excluded from the PMI estimation. Water has a high contribution in biocatalytic processes which has a negative effect on sustainability. The environmental factors are not yet standardized academia- or industry-wide. The PMI metric is nevertheless used by almost 60 % of all chemical companies.

2. Instead, companies and researchers should consider the environmental impact on an entire process via life cycle assessment (LCA). In contrast to the PMI and E factor, LCA comprises all environmental impacts including acquisition of raw materials and catalysts up to disposal or recycling and toxicity issues. LCA is however a very labor-intensive approach, which explains the more frequent use of the PMI or E factor.

3. In the last decades Biocatalysis has developed into a key technology for API synthesis in the pharmaceutical industry and is often referred to as green technology. Using Biocatalysis however imposes reaction conditions also affecting the environment in a certain way, such as the usage of a high amount of water as solvent and a low substrate load, which must be considered for process assessment. Calculating the metrics helps to ascertain whether and to which extent the environmental impact improved during process optimizations e. g. the implementation of biocatalytic steps in drug synthesis.

Challenges

"In order to have faster and broader implementation of enzymatic steps in API syntheses, several challenges are to be addressed"

1. Efficient Enzyme Engineering
2. Time and Efficiency to Develop Biocatalytic Scalable Processes
3. Integration of Biocatalytic Step in Entire Synthetic Landscape of APIs
4. Process Efficiency
5. Interdisciplinarity
6. Compatibility of Enzymes with Process Conditions
7. Systematic Approaches for Multi-enzyme Cascade Reactions
8. Regulatory Aspects

Ideal Synthesis and GC

Dr.Reddy's

Chem. Soc. Rev. **2013**, *42*, 5522-5551

PMI and AE

Chemical Process

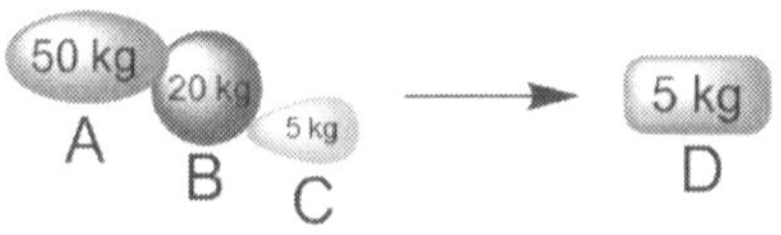

PMI=50+20+5/5=15

PMI=1 is ideal. This is advanced version of E factor

Chemical Process

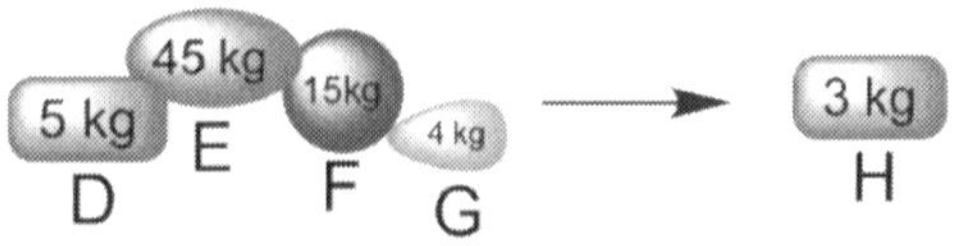

PMI=50(A)+20(B)+5(C)+45(E)+15(F)+4(G)/3=46.3

PMI and AE

Atom Economy of multistep process

$$G \xrightarrow{R} EE \xrightarrow{N} C \xrightarrow{H} E \xrightarrow{M} I \xrightarrow{S} T \xrightarrow{R} Y$$

% Atom Economy for EE=
[MW(EE)/MW(G+R]X100

% Atom Economy for Y=
[MW(Y)/MW(G+R+N+H+M+S+R)X100

Waste and Business

Dr.Reddy's

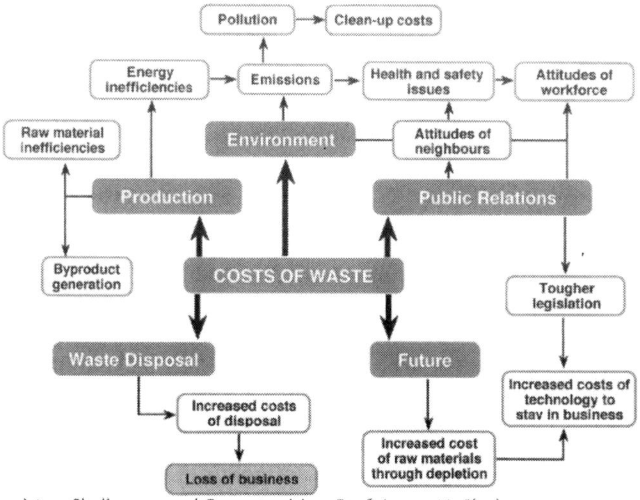

Green Chemistry: Challenges and Opportunities- Prof. James H. Clark

Opportunities

Dr.Reddy's

Environmental impact of some of the chemical industries

Industries	Production Scale (Tons per Year)	E-factor (kg)
Oil Refining	10^6 - 10^8	< 0.1
Bulk Chemicals	10^4 - 10^6	< 1 - 5
Fine Chemicals Industry	10^2 - 10^4	5 - > 50
Pharmaceutical	10 - 10^3	25 - >100

Do we have solution to tame the E-factor or PMI???
-Enzymes can offer some solutions to this

Chemistry in our System

- 2 million proteins
- ~50000 enzymes

Remarkable molecular machinery.........

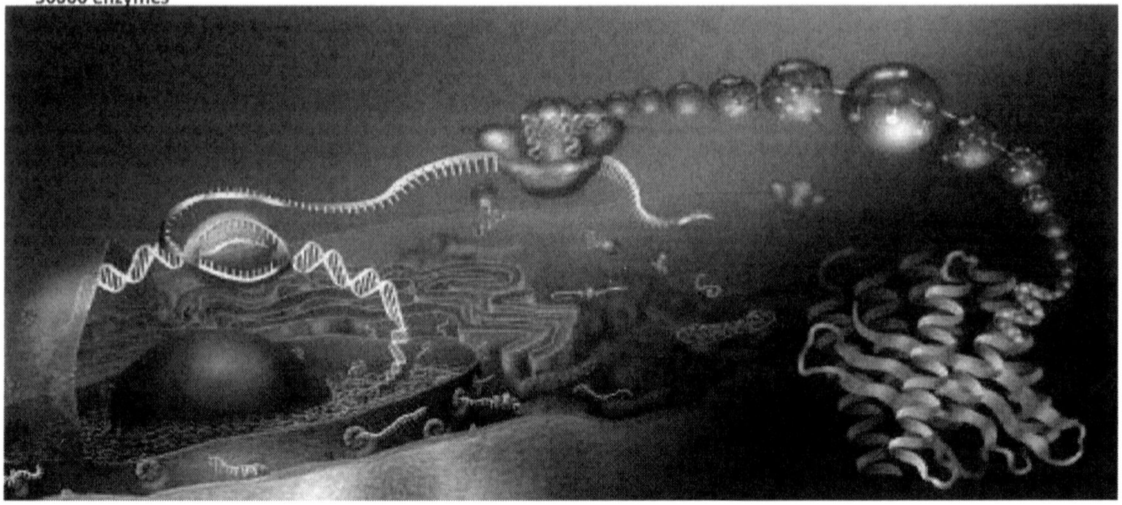

1

Advantages

Dr.Reddy's

- Economics
  - Enzyme contribution to the process is lower
  - It has potential to reduce PMI/E-Factor
- Higher Productivity
  - Higher Yields
  - Concise synthesis route
- Near perfect reaction conditions
  - Mostly ambient reaction condition
  - Usually gets completed in few h
- Higher Selectivity
  - Highest degree of enantioselectivity
  - Other selectivity can also be achieved
- Easy workup
  - Efficient isolation
  - Less number of impurities

Biocatalysis process
- Green
- Atom economical
- Unmatched enantioselectivity

Biocatalysis: Industrial Perspective

Dr.Reddy's

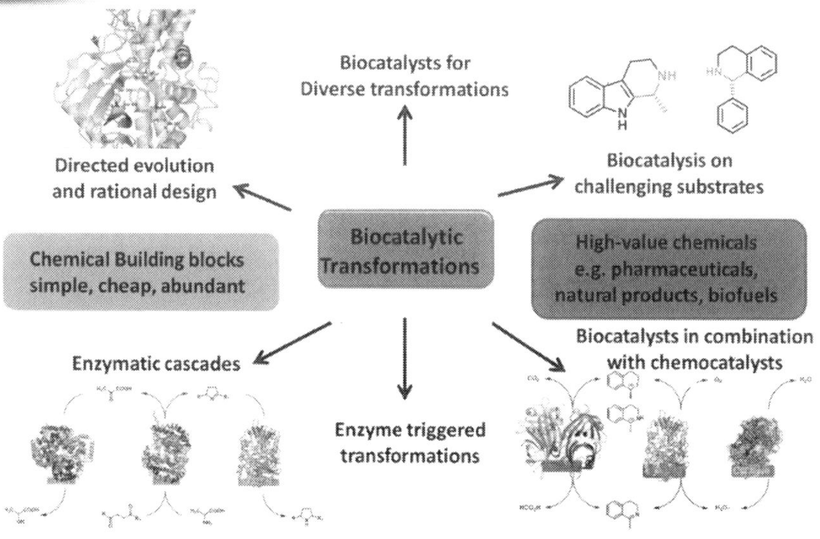

Biocatalysis

Dr.Reddy's

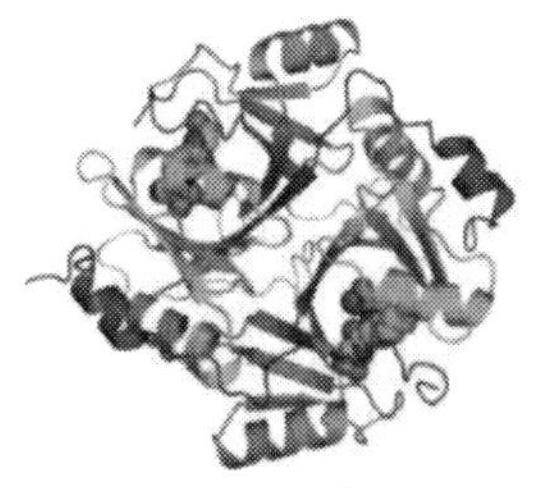

Natural or genetically engineered enzymes

Fermentation based

Whole cell based

Biocatalysis

Dr.Reddy's

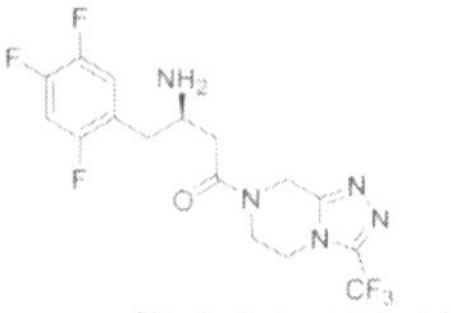

Sitagliptin (*anti-diabetic*)

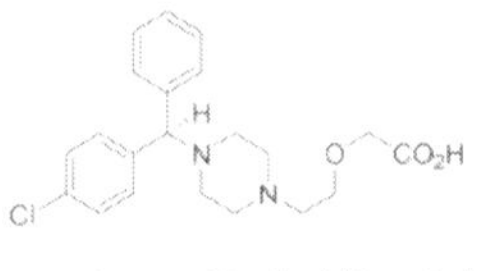

Levocetrizine (*anti-histamine*)

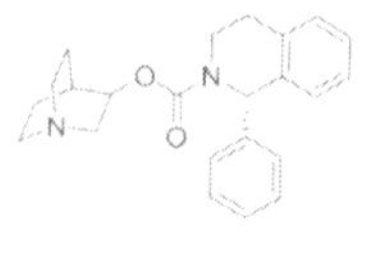

Solifenacin (*anti-muscarinic*)

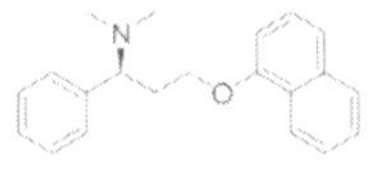

Dapoxetin (*for premature ejaculation*)

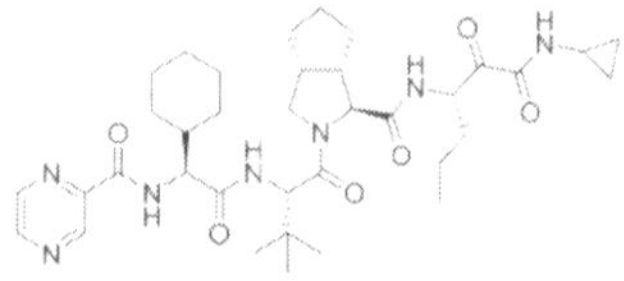

Telaprevir (*anti-HCV*)

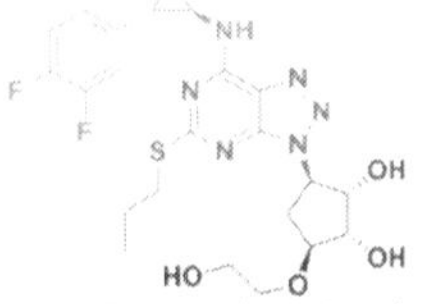

Ticagrelor (*anti-platelet*)

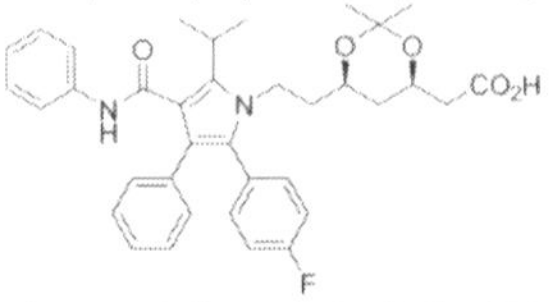

Atorvastatin (*for managing lipid profile*)

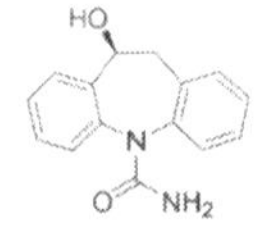

[Esclicarbazepine or (*S*)-Licacarbazepine (*anti-seizure*)

Biocatalysis

Dr.Reddy's

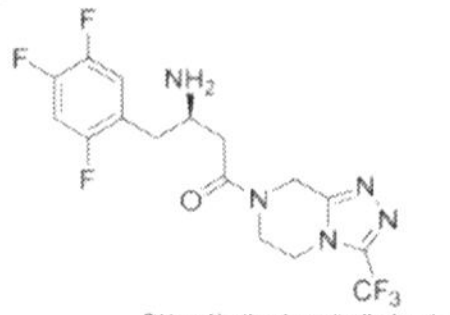

Sitagliptin (*anti-diabetic*)

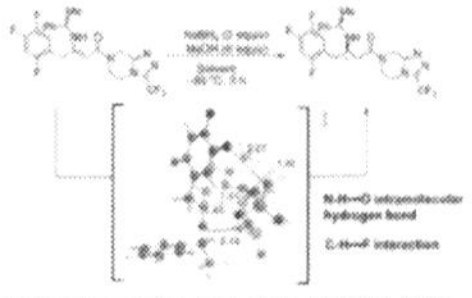

Bandichhor R. et al. Org. Lett. **2015**, *17*, 1742–1745

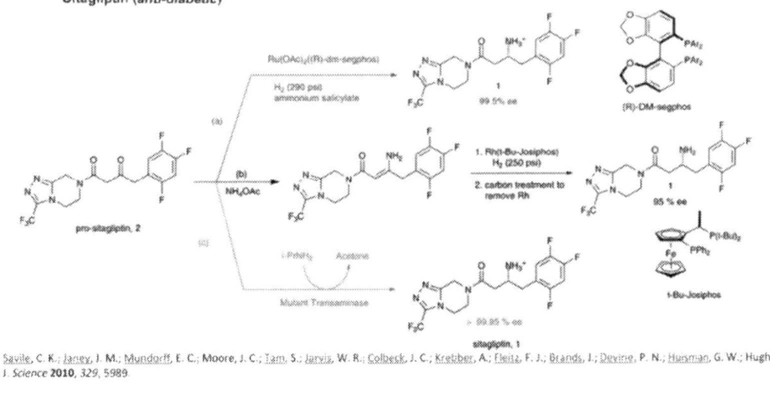

Savile, C. K.; Janey, J. M.; Mundorff, E. C.; Moore, J. C.; Tam, S.; Jarvis, W. R.; Colbeck, J. C.; Krebber, A.; Fleitz, F. J.; Brands, J.; Devine, P. N.; Huisman, G. W.; Hughes, G. J. *Science* **2010**, *329*, 5989.

Biocatalysis

Dr.Reddy's

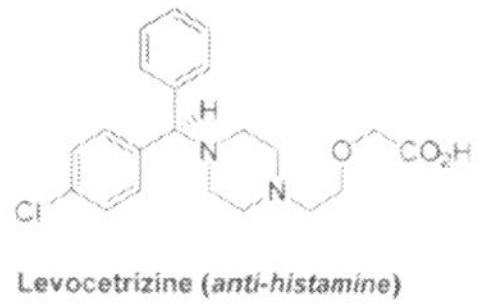

Levocetrizine (*anti-histamine*)

4-chlorobenzhydrylamine:

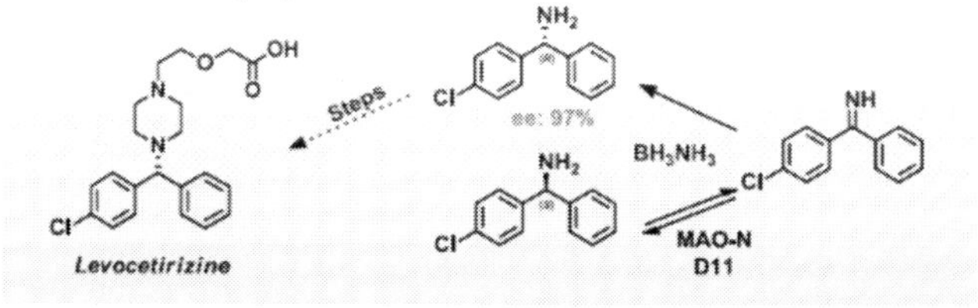

Ghislieri, D.; Green, A.P.; Pontini, M. et al. *J. Am. Chem. Soc.* **2013**, *135*, 10863-10869

Biocatalysis

Dr.Reddy's

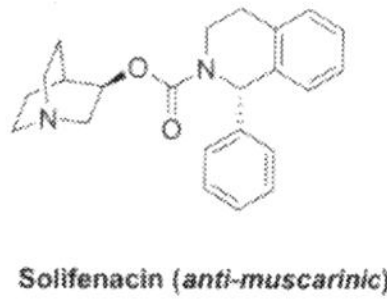

Solifenacin (*anti-muscarinic*)

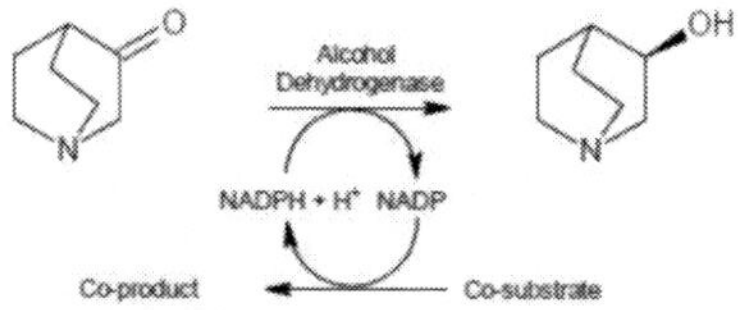

ee >99%

M. T. Reetz, *Angew. Chem. Int. Ed. Engl.*, Bd. 50, 1, 138–174, 2011.

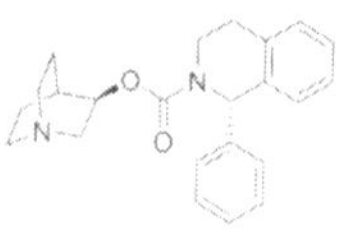

Solifenacin (*anti-muscarinic*)

1-phenyltetrahydroisoquinoline:

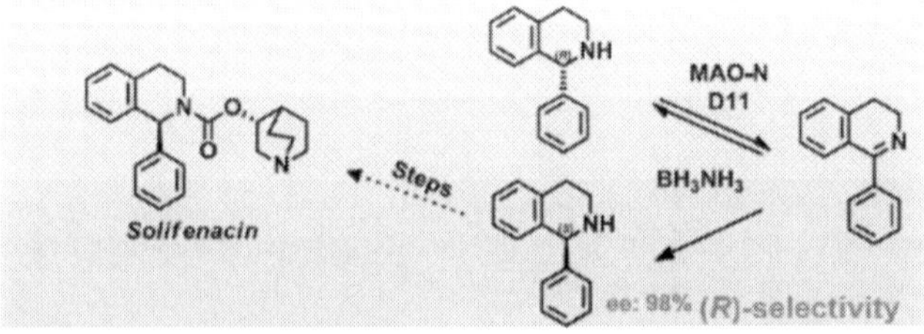

Ghislieri, D.; Green, A.P.; Pontini, M. et al. *J. Am. Chem. Soc.* **2013**, *135*, 10863-10869

Biocatalysis

Dr.Reddy's

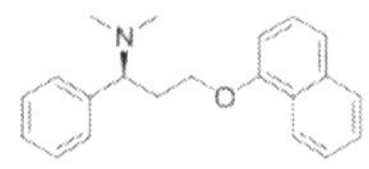

Dapoxetin (*for premature ejacul.*

CAL-A: *C. antartica lipase* A

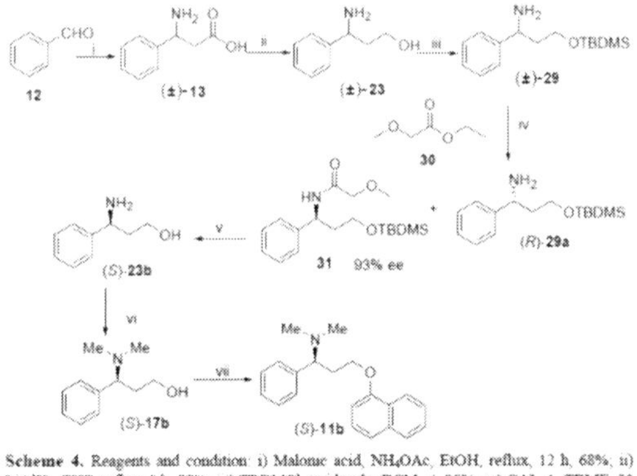

Scheme 4. Reagents and condition: i) Malonic acid, NH_4OAc, EtOH, reflux, 12 h, 68%; ii) $LiAlH_4$, THF, reflux, 3 h, 77%; iii) TBDMCl, imidazole, DCM, rt, 95%; iv) CAL-A, TBME, 30 °C, 250 rpm; v) 6 M HCl, 50 °C, 84%; vi) $(CH_2O)_n$, HCO_2H, reflux, 83%; vii) 1-naphthol, Ph_3P, DEAD, THF, rt, 72%

Torre,O.; Gotor-Fernadaz, V.; Gotor, V. *Tetrahedron: Asymmetry* **2006**, *17*, 860.

Biocatalysis

Dr.Reddy's

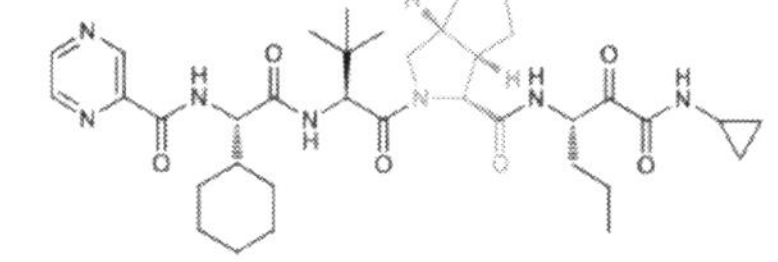

Telaprevir - hepatitis C viral protease inhibitor (phase III)

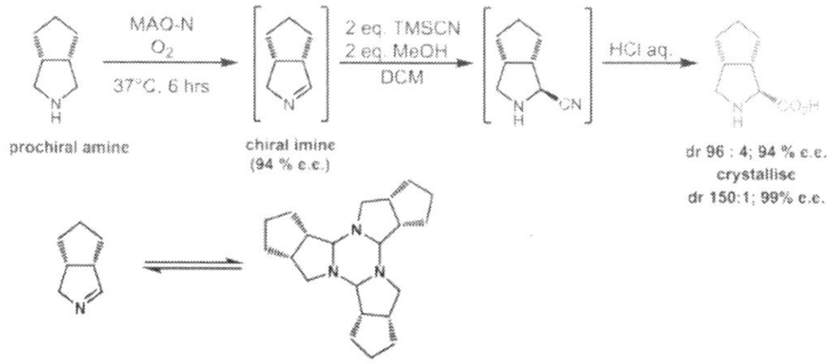

V. Koehler, N.J. Turner *et al.*, *Angew. Chem. Int. Ed.*, 2010, *49*, 2182.

Biocatalysis

Dr.Reddy's

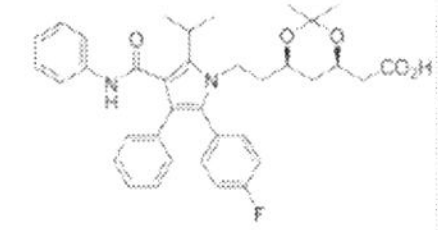

Atorvastatin (*for managing lipid profile*)

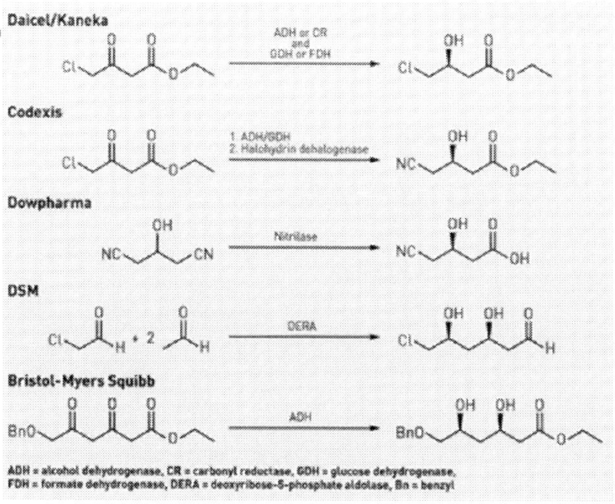

C&EN **2006**, *84*, 26-27

Biocatalysis

Dr.Reddy's

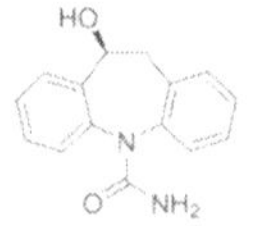

[Eslicarbazepine or (S)-Licarbazepine] (*anti-seizure*)

Asymmetric Reduction of a Key Intermediate of Esclicarbazepine Acetate Using Whole Cell

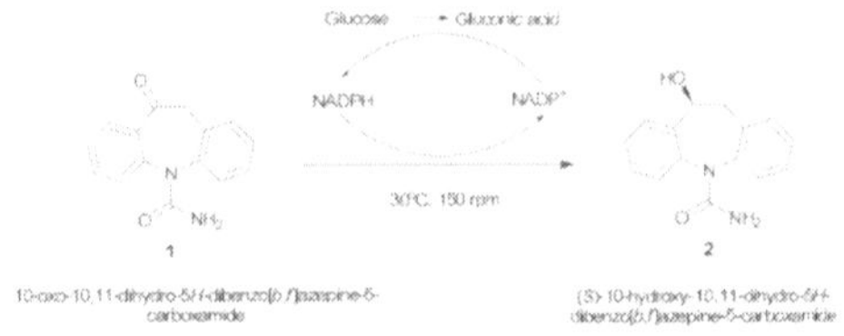

A whole-cell mediated asymmetric reduction of oxcarbazepine **1** to *S*-licarbazepine **2**, a key intermediate in the synthesis of Eslicarbazepine acetate, was achieved using the yeast strain *Pichia methanolica* 103660. The reduction was carried out in a biphasic system consisting of (1 : 1) de-ionized water and hexane. This organism catalyzed the reaction to yield the *S*-enantiomer **2** with excellent conversion (> 98%) and enantiomeric excess (*ee* > 98%) within 24 h of incubation.

Bandichhor et al. *Catalysis Science & Technology* **2012**, *2*, 1602-1605. (*One of the Hot Articles*).

Biocatalysis

Dr.Reddy's

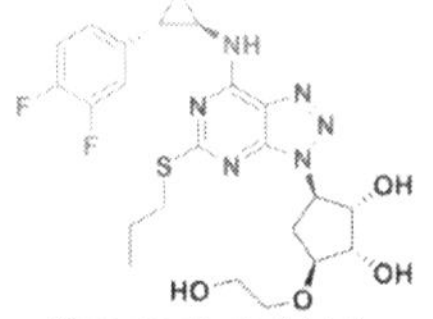

Ticagrelor (*anti-platelet*)

C. parapsilosis has shown complete conversion and excellent enantioselectivity (>98%) at 4.0 g/L substrate concentration and the organism is robust enough to grow on natural media. Our whole-cell catalyzed system has the advantage of not requiring external cofactors or cofactor recycling systems, thus reducing cost.

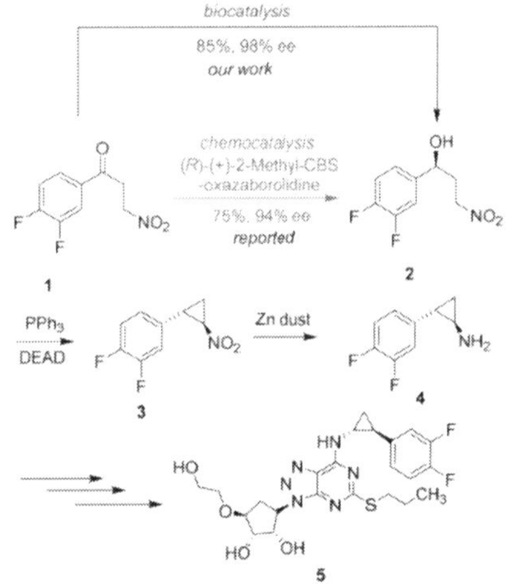

Bandichhor et al. ***RSC Adv.***, **2016**, *6*, 35086-35090

Biocatalysis in the Synthesis of Naproxen

Old Yellow Enzymes (OYEs) class: NerA (*Agrobacterium radiobacter*) and XenA (*Pseudomonas putida*)

FMN: Flavin mononucleotide

Ene reductases

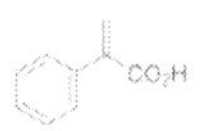

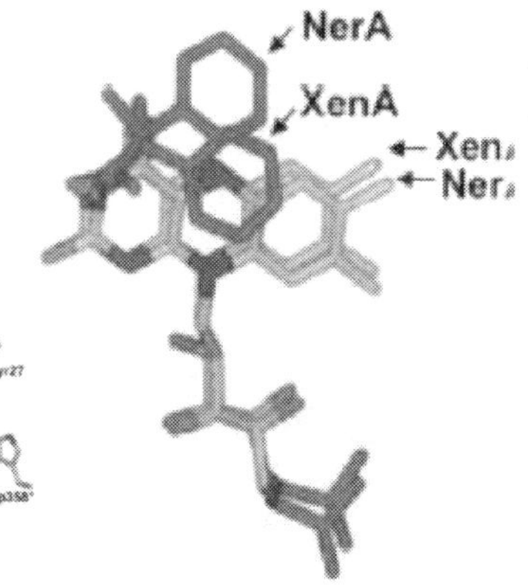

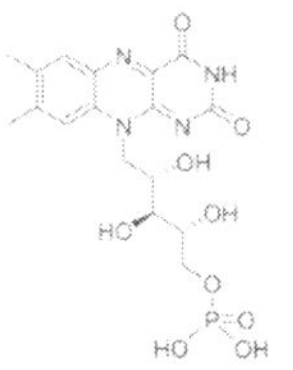

Works as prosthetic group

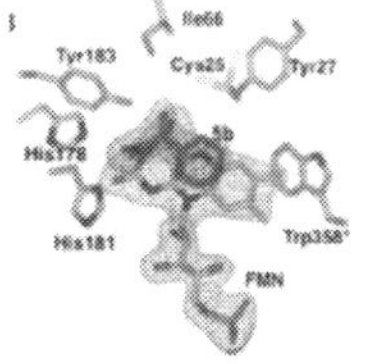

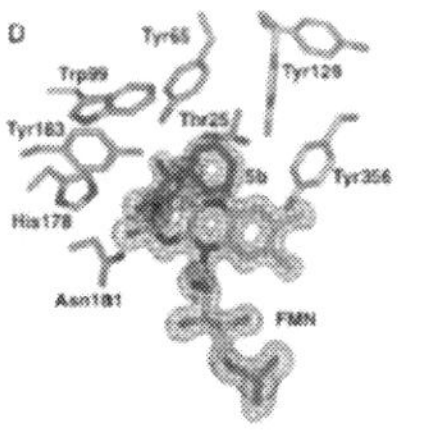

Biocatalysis in the Synthesis of Naproxen

Dr.Reddy's

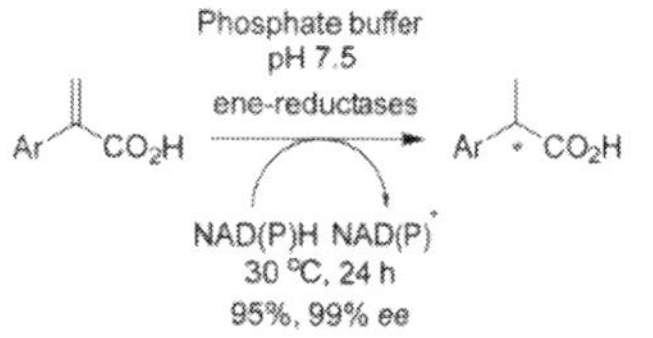

Table 1 Biotransformations of ene-reductases with profens **3b** and **5a-c**.

Substrate	Product	Enzyme	Conv. (%)[a]	*ee* (%)[b]
3a	(*R*)-**4a** 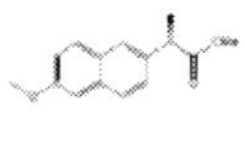	XenA	57	>99 (*R*)
		GYE	24	>99 (*R*)
		LeOPR1	21	>99 (*R*)
		TOYE	17	>99 (*R*)
		PETNR	9	>99 (*R*)
		NerA	3	>99 (*R*)
5a	(*R*)-**1a** 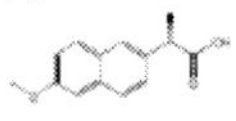	XenA	94	>99 (*R*)
		GYE	95	>99 (*R*)
5b	(*R*)-**5c** 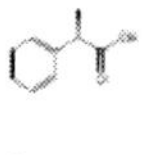	XenA	97[c]	>99 (*R*)
		GYE	44[c]	>99 (*R*)
		YqjM[d]	95[d] (ref 7)	>99 (*R*)
5c	**1c** 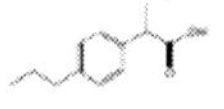	XenA	7[c,d]	N/D
		GYE	27[c,d]	N/D

Bandichhor et al. *Org. Biomol. Chem.* **2017**, *15*, 4440-4448

Biocatalysis

Dr.Reddy's

Multienzyme cascade reactions are the new trend having capability to totally replace the chemical steps

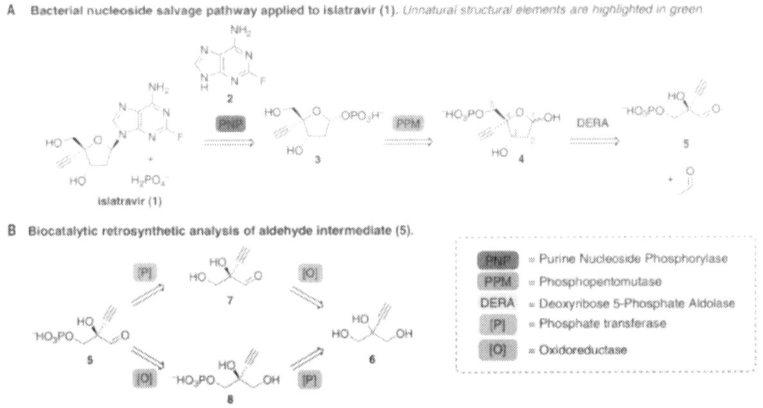

Huffman et al. **Science., 2019**, *366*, 1255-1259

Biocatalysis

Dr.Reddy's

Successful synthesis of various Drugs using Multienzyme cascades

Enzyme	Source Organism	Modification	Biocatalyst	Substrate	Product	Process Performance
IRED; KRED	Diverse	Directed evolution (one enzyme)	Isolated enzymes	Amine tranylcypromine sulfate, alcohol precursor	Protected GSK-2879552	5 g scale with 48.3% yield, 99.5%-ee, 97.9% purity
Nov435; MTR kinase; UP; AcK; catalase; POX	Diverse	Directed evolution (two enzymes)	Isolated enzymes	Ribose, uracil	Molnupiravir	69% overall yield
GOase; HRP; catalase; PanK; AcK; PNP; PPM; DERA; SP	Diverse	Directed evolution (five enzymes)	Isolated enzymes	2-Ethynylglycerol	Islatravir	51% overall yield
ThiM; IPK; IDI; GPPS; ADK; AcKA; PTA; MdcA; AAE3; OLS; OAC; CBGA synthase	Diverse	One engineered enzyme by Rosetta design [151]	Isolated enzymes	Isoprenol, acetylphosphate, malonate, butyrate/hexanoate	Cannabigerolic acid (CBGA); cannabigerovarinic acid (CBGVA)	480 mg L^{-1} (CBGA), 580 mg L^{-1} (CBGVA)
ADK; 2 PPK2; cGAS	Diverse	-	Isolated enzymes	Adenosine, GTP	2′3′-cGAMP	0.08 mol per mol adenosine
GK; AK; AcK; cGAS	Diverse	Directed evolution (four enzymes)	Isolated enzymes	Nucleotide monothiophosphates	MK-1454	62% isolated yield

Kinner et al. **Biomedicines., 2022,** *10*, 964-995

Thank You

MERCK

Development of a Sustainable Biocatalytic Process for the Manufacture of an Amino-Alcohol

Jacob H. Forstater

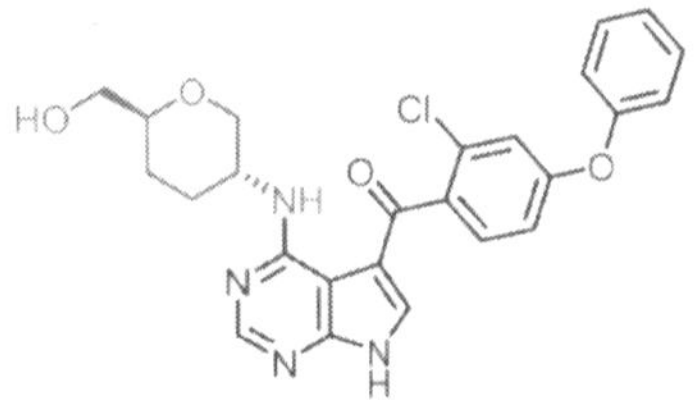

Nemtabrutinib (MK-1026)

AiChE PDS 2022
Philadelphia, PA

June 09, 2022

MK-1026 – A Reversible BTK Inhibitor

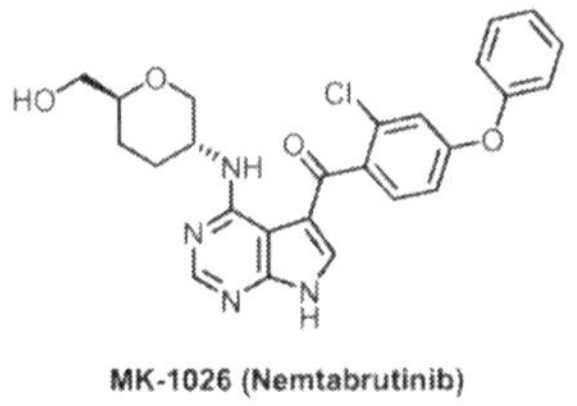

MK-1026 (Nemtabrutinib)

ARQULE®

- Indication: Treatment of hematologic malignancies such as chronic lymphocytic leukemia (CLL) and Non-Hodgkin's Lymphoma
- Arqule acquired in January 2020
- Mechanism: ***Reversible*** inhibitor of Bruton's tyrosine kinase (BTK)
- Clinical phase: Phase 2

2

A reversible BTK Inhibitor to Address Treatment Resistance

- Treatment of hematologic malignancies such as chronic lymphocytic leukemia (CLL) and Non-Hodgkin's Lymphoma

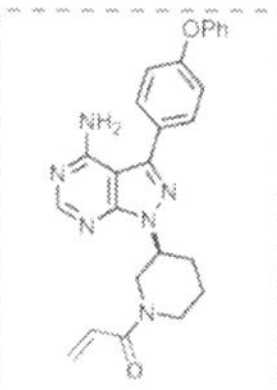

Ibrutinib, current standard of care

Binding mode of **ibrutinib** with BTK

Irreversible BTK inhibitors covalently bind to cysteine in position 481. A **cysteine to serine mutation** (C481S) prevents this interaction.

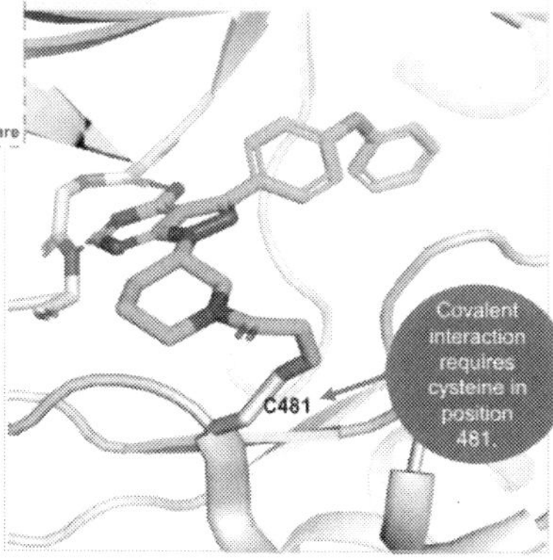

A reversible BTK Inhibitor to Address Treatment Resistance

- Treatment of hematologic malignancies such as chronic lymphocytic leukemia (CLL) and Non-Hodgkin's Lymphoma

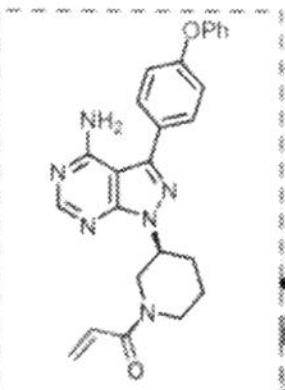

Ibrutinib, current standard of care

Binding mode of **ibrutinib** with BTK

Irreversible BTK inhibitors covalently bind to cysteine in position 481. A **cysteine to serine mutation** (C481S) prevents this interaction.

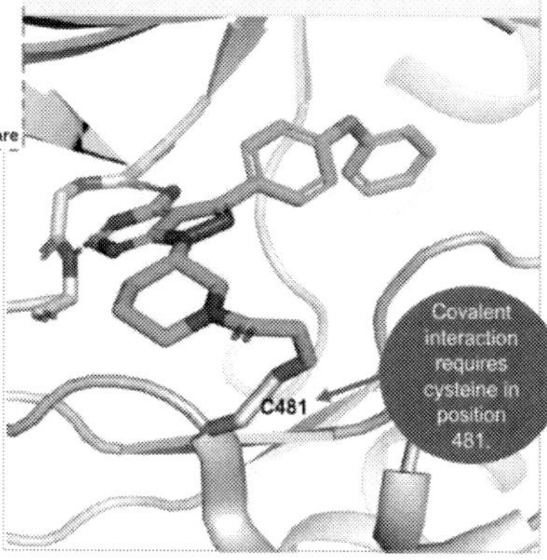

Crystal structure of **MK-1026** with BTK

MK-1026 was **specifically designed** to inhibit BTK regardless of the amino acid present in position 481

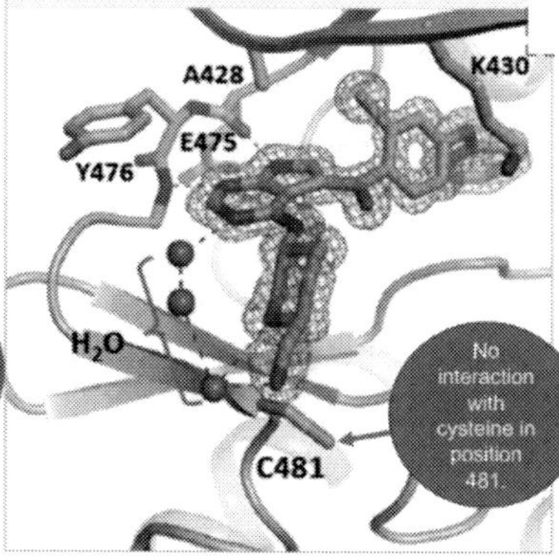

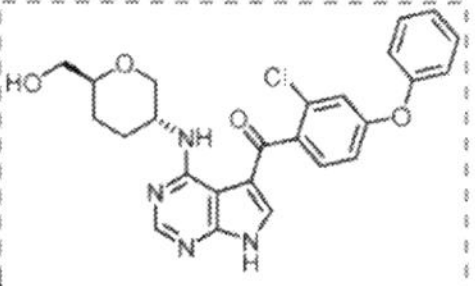

MK-1026 (Nemtabrutinib)

- Original Synthesis: *Eur J. Org. Chem.* **2003**, 2418.

MK-1026 – Synthetic Chemistry Priorities

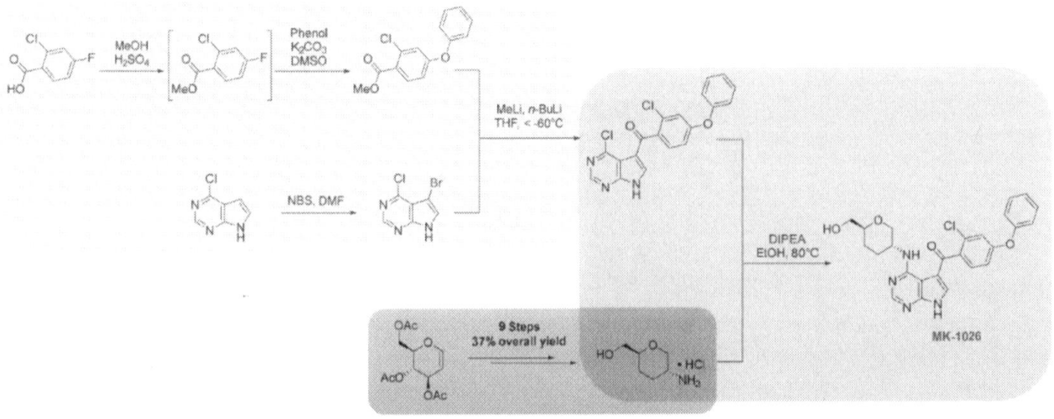

Key Goal for Route Development:
Development of a more efficient synthesis of the amino alcohol fragment

5

Cyrene: Converting a Solvent into a Key Intermediate

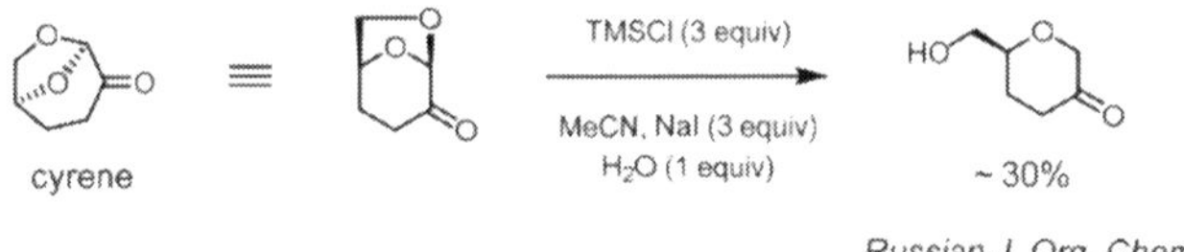

Russian J. Org. Chem. **2015**, *51*, 569

- Prepared by acid-catalyzed pyrolysis of biomass
- Green and renewable
- Typically used as a solvent, alternative to NMP and DMF

Reductive amination (bio- or chemocatalytic)

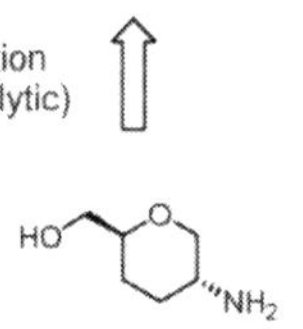

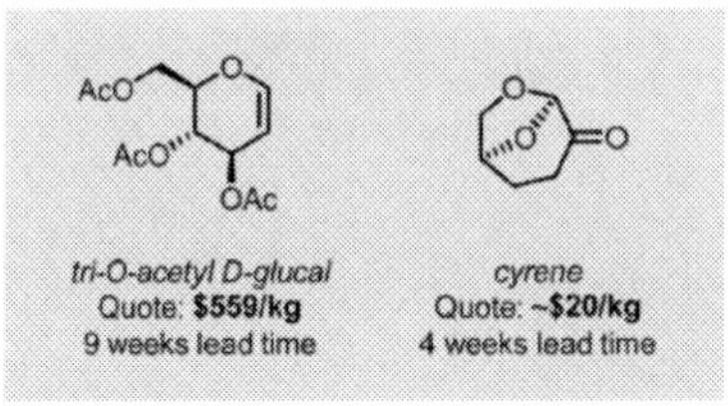

Cyrene™

CAS No.: 53716-82-8
Alternative for: NMP, DMF, and sulfolane

Features & benefits

Cyrene™ is a new dipolar aprotic alternative to common REACH-restricted solvents, such as N-methyl-2-pyrrolidone (NMP). One of the first true greener solvents, it is produced from renewable resources, and safe for end of life disposal, decomposing into CO_2 and H_2O. Cyrene™ is also safer to handle, with no mutagenic or genotoxic concerns.

- 99% biodegradation in 28 days
- Stable during incineration
- Not mutagenic or genotoxic

J. E. Camp, *ChemSusChem* **2018**, *11*, 3048.

7

Routes Investigated from Cyrene to Amino Alcohol

Ketone Stability

- Intermediate unstable in end-of-reaction stream, neat, or basic conditions
- Isolated in a dilute brine stream

Transamination/Reductive Amination

- Enzyme evolution complicated by ketone instability
- High enzyme loads needed due to the high salt concentration of the ketone stream
- Chemical routes result in <3:1 d.r. of desired amine

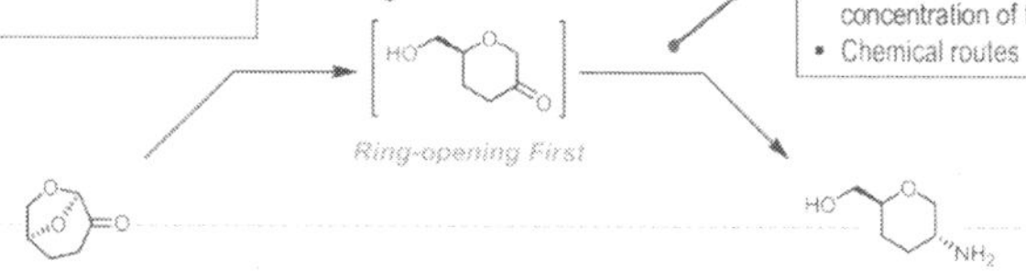

8

Routes Investigated from Cyrene to Amino Alcohol

Ketone Stability

- Intermediate unstable in end-of-reaction stream, neat, or basic conditions
- Isolated in a dilute brine stream

Transamination/Reductive Amination

- Enzyme evolution complicated by ketone instability
- High enzyme loads needed due to the high salt concentration of the ketone stream
- Chemical routes result in <3:1 d.r. of desired amine

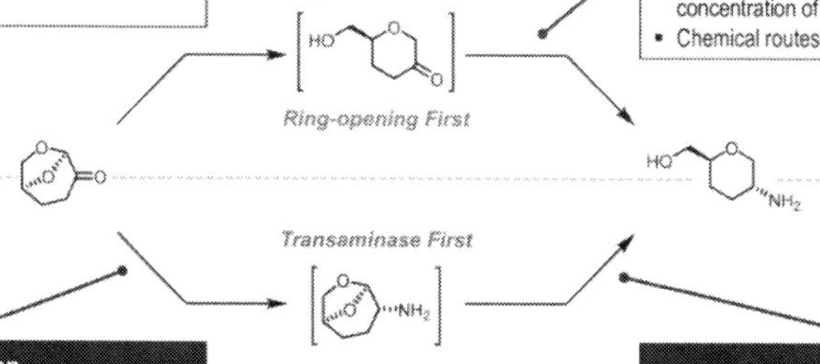

Transamination

- Enzyme evolution using a commodity chemical
- Access to stable, crystalline amine salts with high d.r.

Acetal-Opening

- Protecting group free ring-opening conditions identified

"Transaminase first" route selected for development

9

Transamination of Cyrene

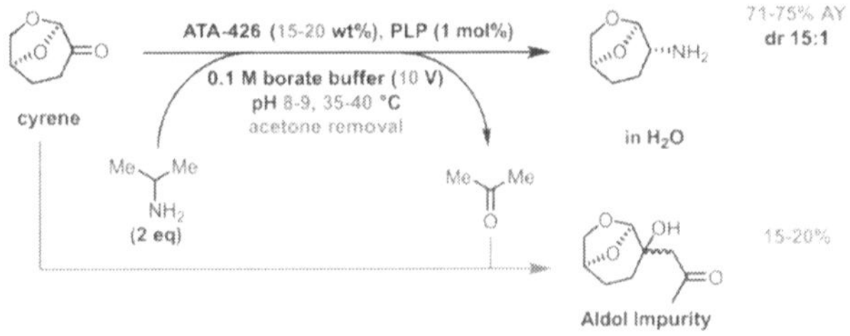

- ATA-426 performs the transamination of cyrene favoring desired isomer (15:1 dr)
- Acetone byproduct generates large quantities of aldol byproducts (15-20%)

Savile, C. K. et al. *Science* **2010**, *329*, 305.

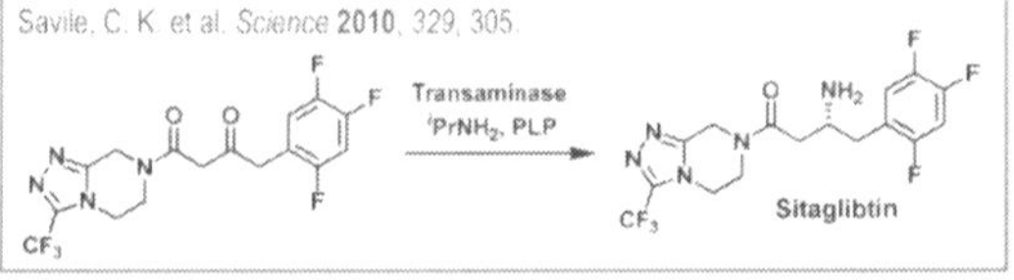

Yasuda, N. et al. *Org. Process Res Dev.* **2017**, *21*, 1851.

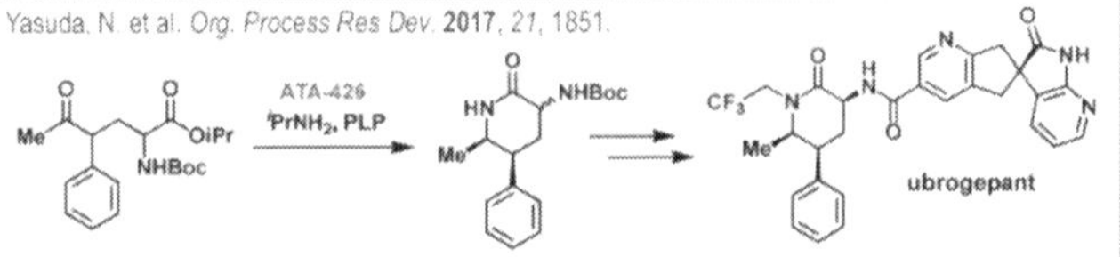

ATA-426 is available from CODEXIS

10

Transamination of Cyrene

- Combination of N_2 **sweep and vacuum** controls aldol impurity to <2%, enables cyrene amine yield >90%
- Isolation of cyrene amine from aqueous stream is very challenging
    - Boc protection, salting out, and direct crystallization approaches developed, but drawbacks to each
    - Challenging **protein removal**

Can enzyme immobilization enable transamination in organic solvent?

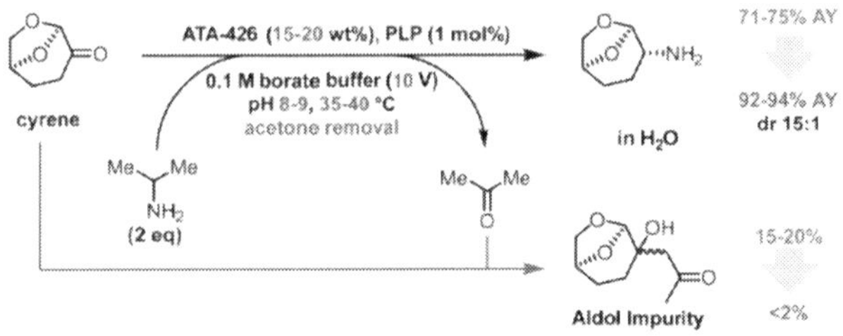

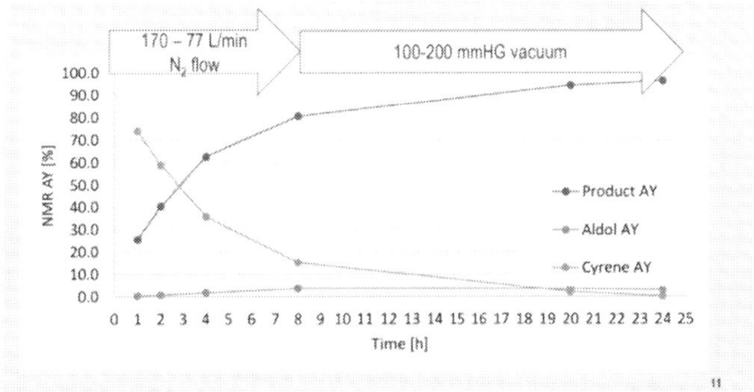

11

Enzyme Immobilization by Adsorption (*or how to minimize enzyme dehydration)*

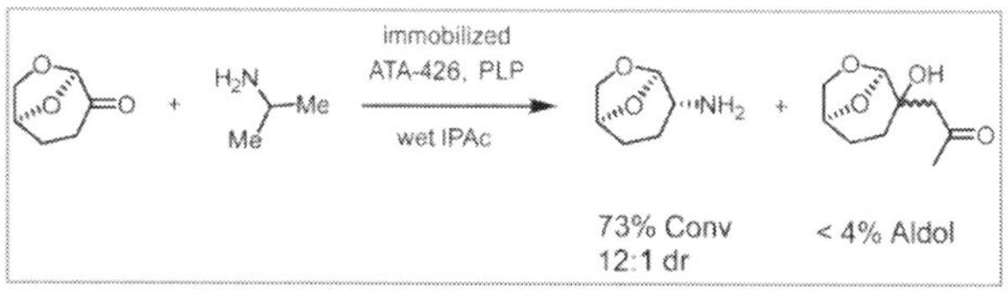

Data-Rich Experimentation: Rapid Reaction Optimization in Batch

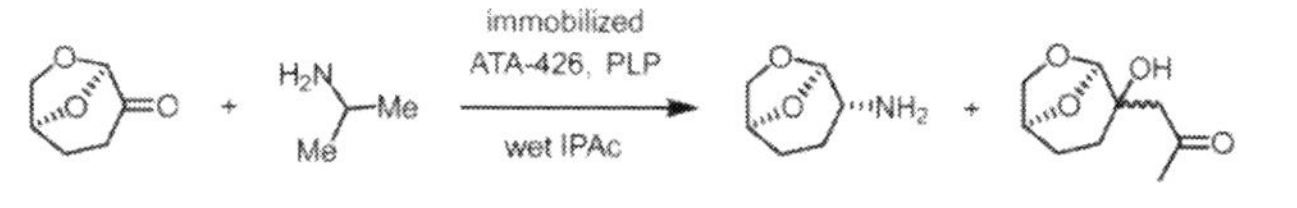

- Performed high-throughput optimization using DoE approach with batch immobilized reactions
- Results further supported by kinetic studies
- Optimized conditions: 100 g/L cyrene and 1.25 equiv. isopropylamine, 60 °C

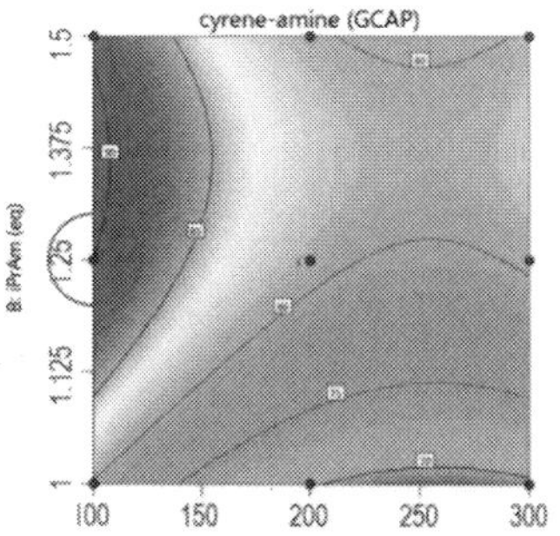

Product purity/yield improves with lower Cyrene concentration

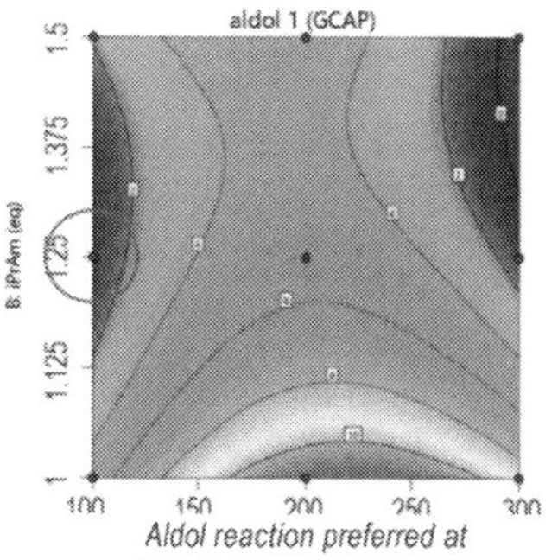

Aldol reaction preferred at high Cyrene concentrations. 1st order in [Cyrene] and [acetone].

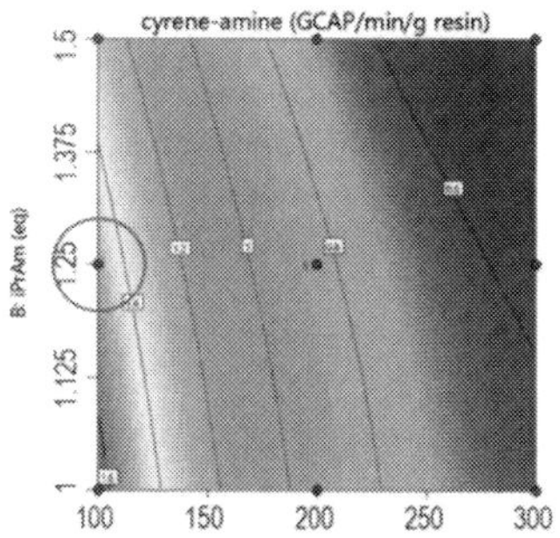

Transamination is faster at low $^{i}PrNH_2$ concentrations.

13

Data-Rich Experimentation: Rapid Reaction Optimization in Batch

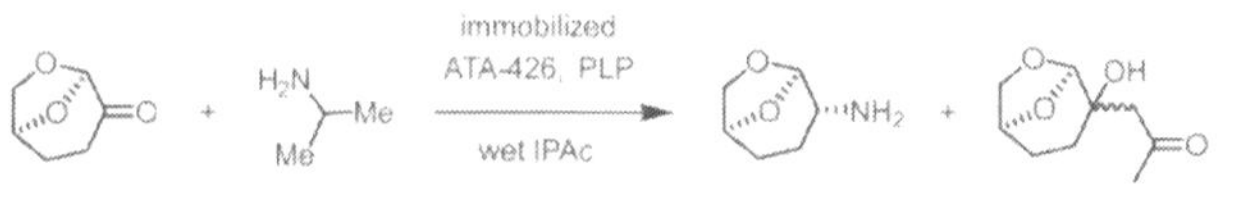

- Performed high-throughput optimization using DoE approach with batch immobilized reactions
- Results further supported by kinetic studies
- Optimized conditions: 100 g/L cyrene and 1.25 equiv. isopropylamine, 60 °C

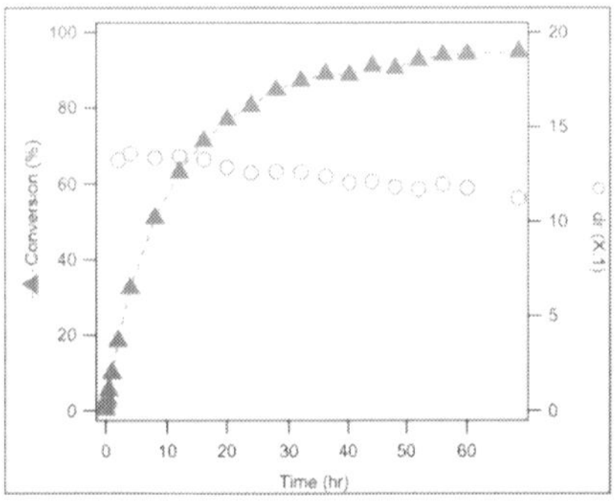

Options for Cyrene Transamination Reaction

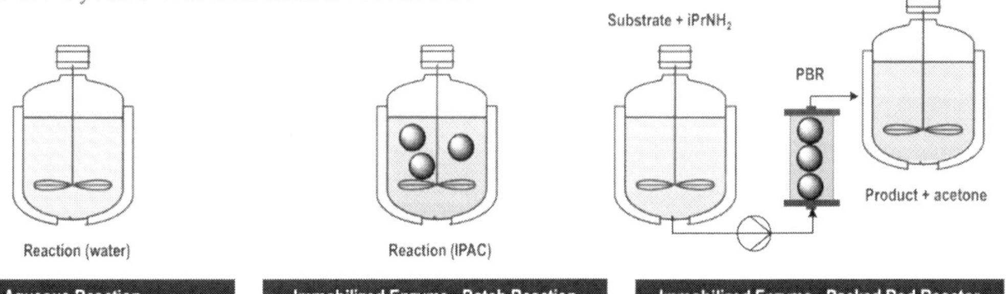

Aqueous Reaction	Immobilized Enzyme - Batch Reaction	Immobilized Enzyme - Packed Bed Reactor
• Acetone removal (N_2 stream, vacuum) poses challenges for scale-up • Requires in situ pH adjustment • Challenging isolation of water-soluble product • Challenging protein removal	• Simplified isolation from an organic stream • Lower aldol formation (<4%) • No need to remove acetone • Multiple unit operations to immobilize enzyme and condition resin, risks on scale	• Reduced handling of resin by conditioning and reaction in a single vessel • Washing/conditioning of resin more consistent • Improved reaction efficiency (higher enzyme concentration vs. substrate in column)

Selected immobilized enzyme / flow process for development

15

Dynamic Flow Platform for Packed-Bed Reactor Experiments

How to optimize a continuous process?

Traditional Flow Experiment
- Analogous to reaction time in batch
- Each data point (steady-state condition) established by flowing reaction solution over 3 residence times
- **Time and material intensive**

Dynamic Flow Experiment
- Dynamic ramping of flow rate to interrogate range of residence times in a single experiment
- Full kinetic profile of flow conditions in a single day
- **Saves time and materials**

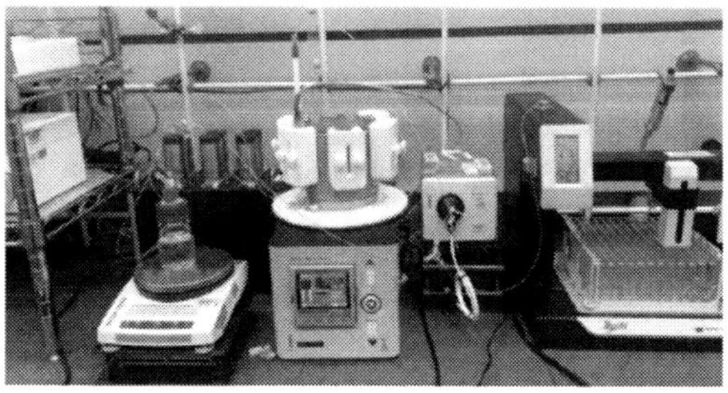

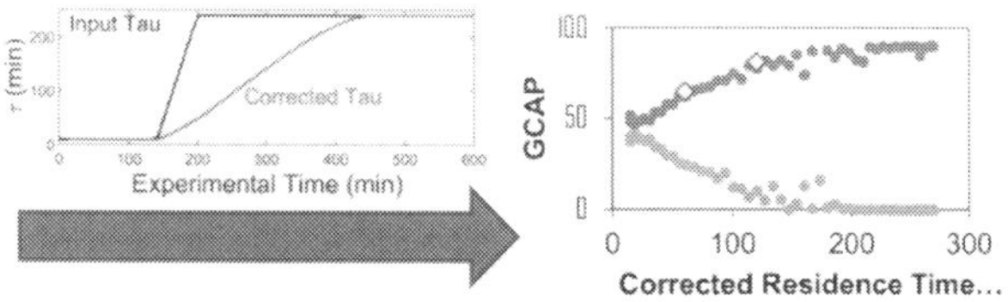

Rapid collection of data under different reaction conditions across broad spectrum of residence times

Wyvratt, B. M.; McMullen, J. P.; Grosser, S. T. *React. Chem. Eng.* **2019**, *4*, 1637-1645.

16

Optimization of Immobilized Flow Process

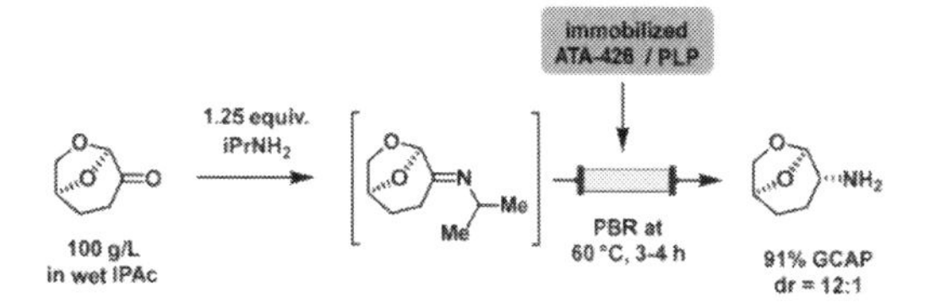

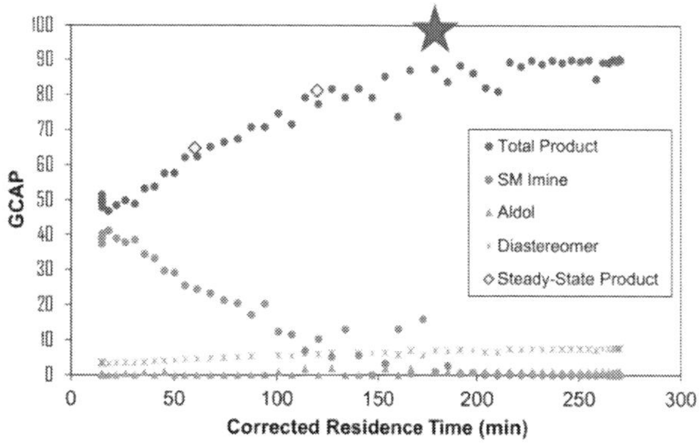

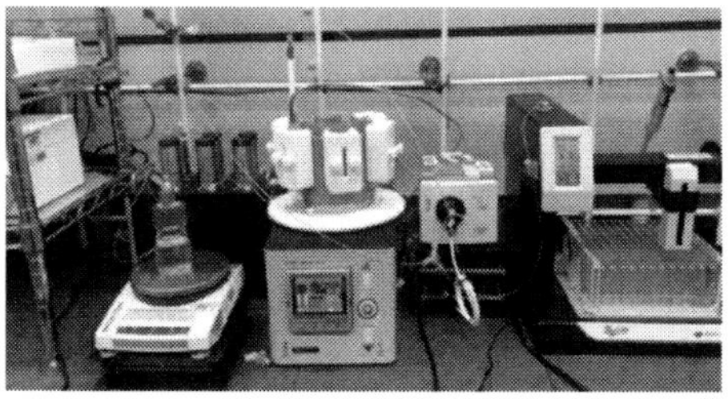

- 3 h residence time is optimal,
- 100 g/L cyrene, 1.25 eq. $iPrNH_2$, 1.25 eq $iPrNH_2$, 100g/L .
- Rapid optimization allowed to go from g to kg in 3 month.
- Processed 25 kg of cyrene in a 20L PBR in 5 days yielding 75% yield of amine product in 17:1 dr

17

Crystallization of Cyrene Amine TsOH Salt

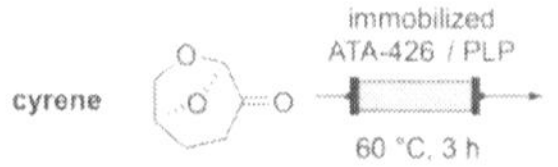

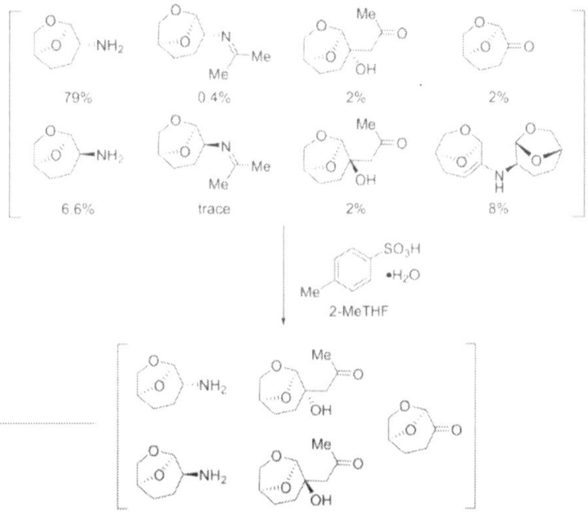

Benefits of Tosylate Salt Crystallization

- Tosic acid hydrate hydrolyzes all imines in solution
- Salt formation completely rejects aldol impurities
- Provides increase in d.r.

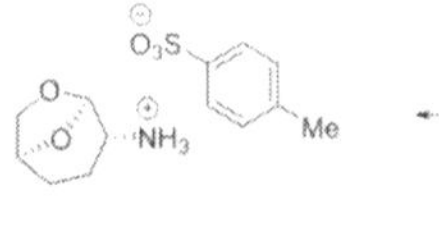

95% IY from amine
dr > 30:1

Robust crystallization provides quality product from streams of varying conversion

INVENTING FOR LIFE

How do we make the process more efficient?

INVENTING FOR LIFE

Improving Transamination through Evolution of ATA-426

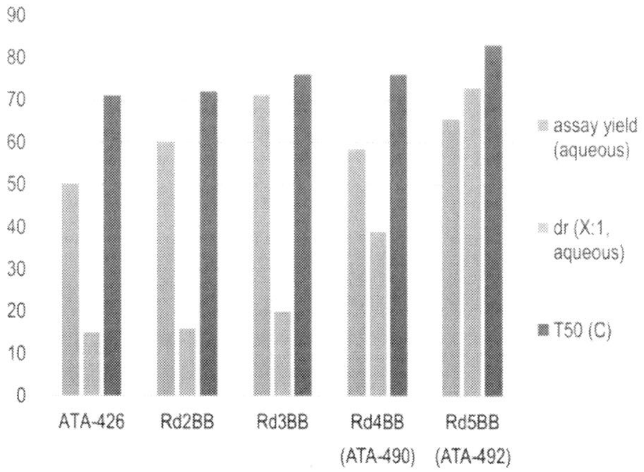

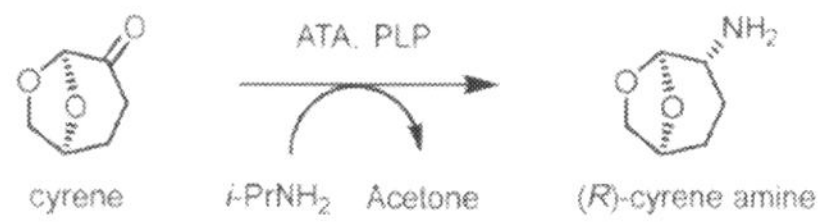

Protein Engineering Strategy

- Site saturation libraries screened for improvements in activity, selectivity, stability
- Combinatorial libraries built using beneficial mutations for rapid evolution
- Evolved under aqueous conditions, and assayed for impacts on immobilized reactions
- 4 rounds of evolution provided a transaminase with improved activity, selectivity, and thermostability

Aqueous reactions; Conditions: 75 g/L cyrene, 5 wt% transaminase, 1 g/L PLP, 0.1 M borate, 1 M $iPrNH_2$, pH 10.5, 45 °C

Diastereoselectivity of Immobilized Transaminase Reactions

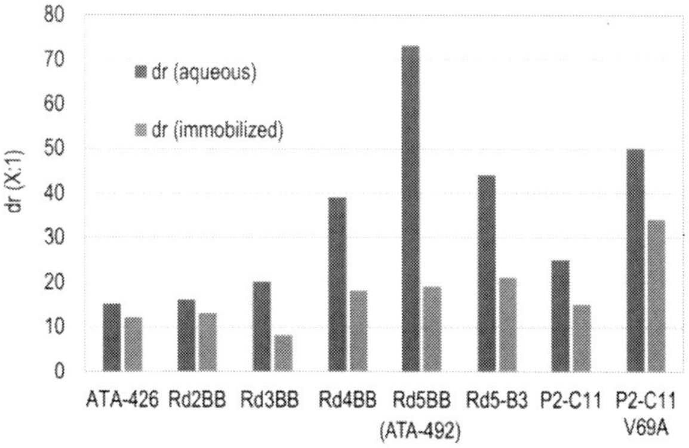

T_{50} (°C)	*69°C*	*72°C*	*76°C*	*76°C*	*83°C*	*78°C*	*68°C*	*68°C*

Reaction conditions: 100 mg/mL cyrene, 100 wt% immobilized enzyme/wet resin, 1.25 equiv. isopropylamine, 55 °C

ATA variants generally display *lower dr under immobilized conditions*

Factors that influence dr: temperature, solvent, time, enzyme variant

Factors that *do not* influence dr: pH of immobilization, ratio of enzyme to resin, resin chemistry, resin pore size

P2-C11 and P2-C11 V69A: variants from a "side path" evolution aimed at optimizing dr when immobilized; variants not selected for development due to lower activity

21

Solvent Impacts on Immobilized Transamination Reaction

- Use of **water-saturated MeTHF** as solvent improves diastereoselectivity of transamination
- Allows for single-solvent transamination reaction and crystallization (process simplification)

Immobilized enzyme flow reactions: ATA-490 immobilized on HP2MGL resin, in IPAc and MeTHF, 90 min residence time:

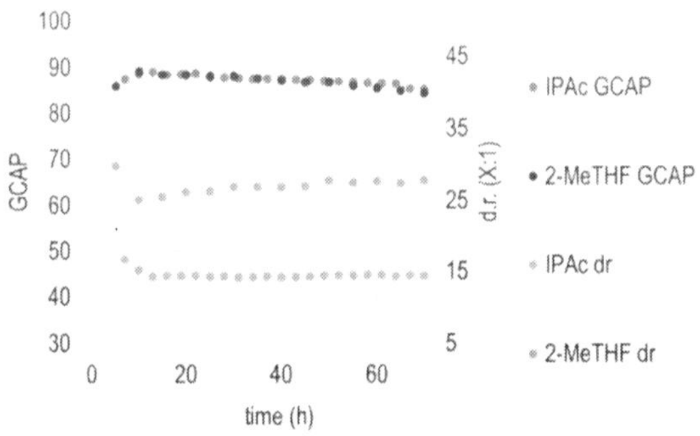

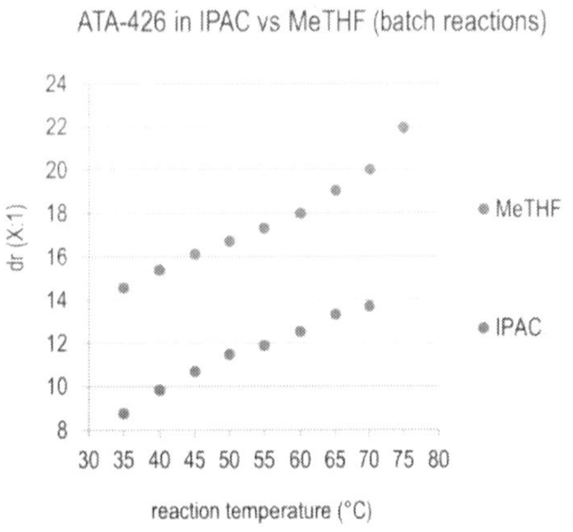

Immobilized Transaminase: Resin Chemistry Impacts Activity

Requirements
- Porous
- Highly Crosslinked
- Process Compatible
- Commercially Available

Variables Investigated
- Bead chemistry
- Bead functionalization and linker
- Pore size
- Particle size

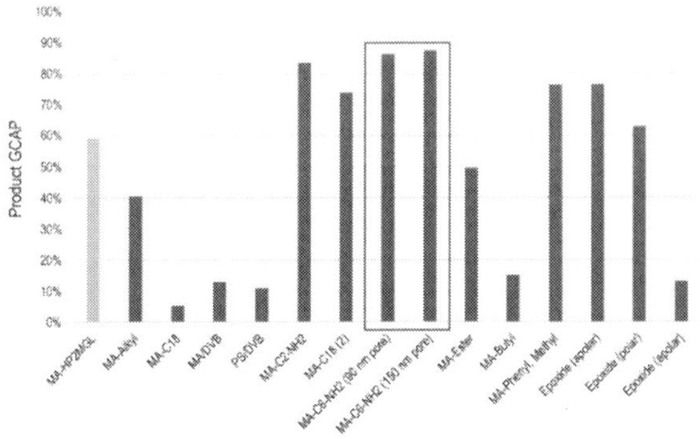

Resins tested with ATA-492 in 2-MeTHF

- Identified an **amine-derivatized resin (ECR8415 from Purolite)** that provides activity improvement
- Amine-functionalized resins preferentially adsorb ATA over host-cell proteins; higher activity achieved in part through increased transaminase loading
- Potential change to immobilization mode
- dr not significantly impacted by changes to resin chemistry, hydrophobicity, or linker
- Faster rate of adsorption on ECR; **immobilization time reduced** from 48 h to 18 h

ECR8415 resin:

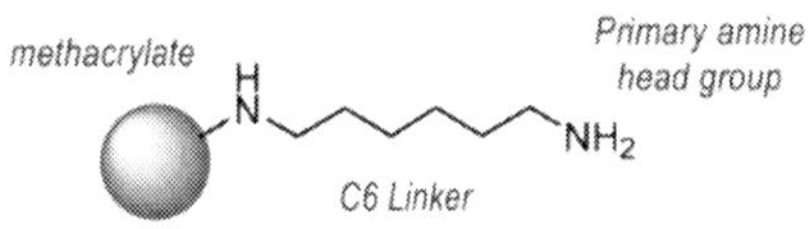

Immobilization of Evolved Transaminase

- Improvements obtained through evolution under aqueous conditions translate to immobilized enzyme flow process
- Good stability and reduced residence time (RT) (From 180 min to 90 min)

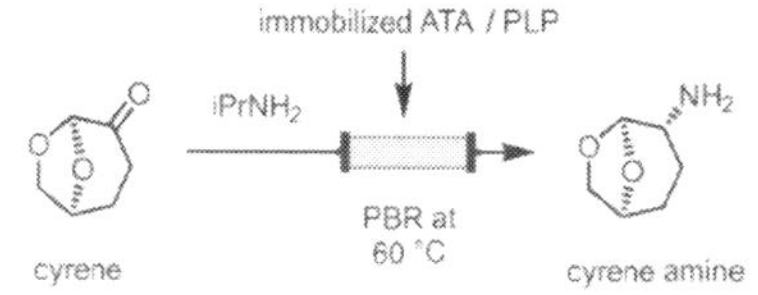

Dynamic Flow Experiments

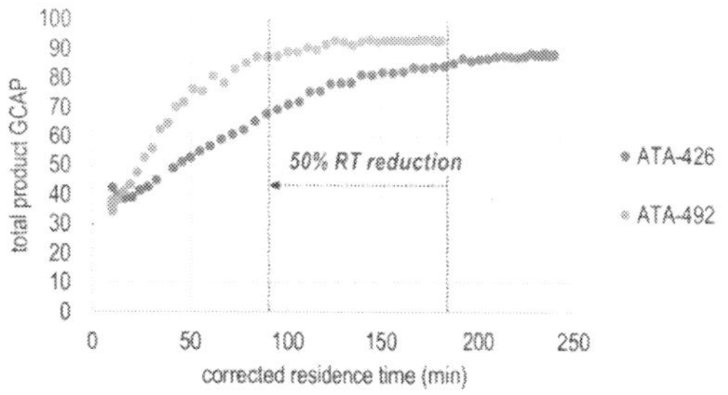

Stability Flow Experiment: ATA-492, 90 min residence time

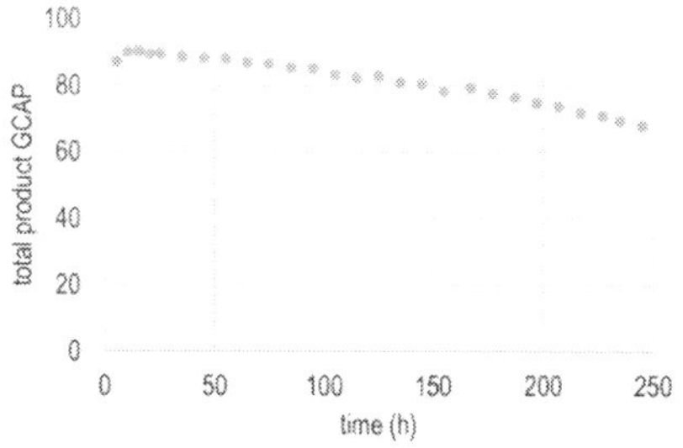

Conditions: 100 g/L cyrene, ATA-426 or ATA-492 immobilized on HP2MGL, 1.25 equiv. $iPrNH_2$, 60 °C

24

Immobilized Transaminase Process Improvements

Improved Enzyme

Change Solvent

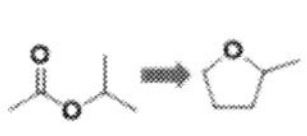

New Resin

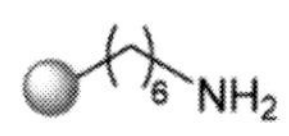

Residence time

3 h to 45 min

Comparing Process Improvements, Dynamic Flow Experiments

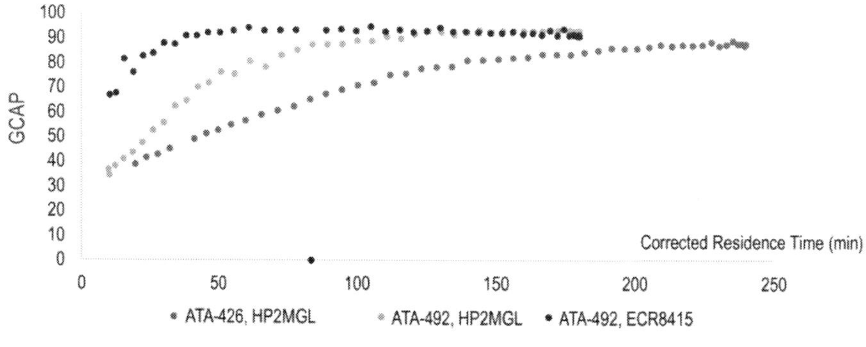

- Combining evolved transaminase and ECR resin leads to a 4-fold reduction in residence time
- Stability of immobilized enzyme is maintained on ECR resin
- Activity improvements allow a smaller column size to produce the same (or greater) amount of product
- Improved process efficiency – more sustainable process

25

Immobilized Transaminase Flow Process: "Gen 2"

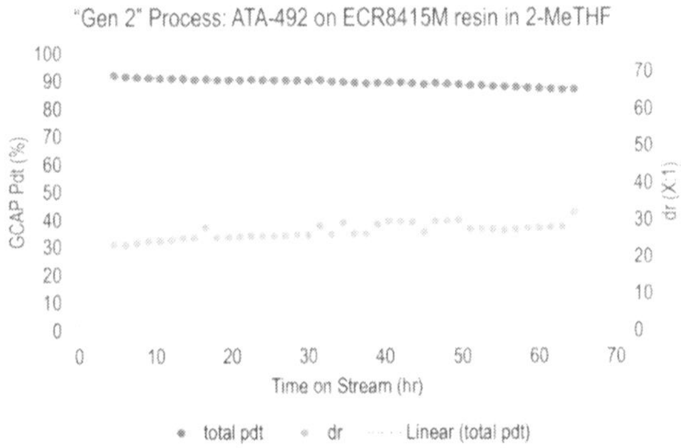

- 84% AY cyrene amine, **45 minute** residence time in 1 L column (Total Batch: 3.1 kg)
- Improved crystallization: addition of water improves rejection of undesired diastereomer
- Isolation as TsOH upgrades from ~88% GCAP to 99.8% GCAP with dr of ~470:1

26

Immobilized Transaminase Process Improvements

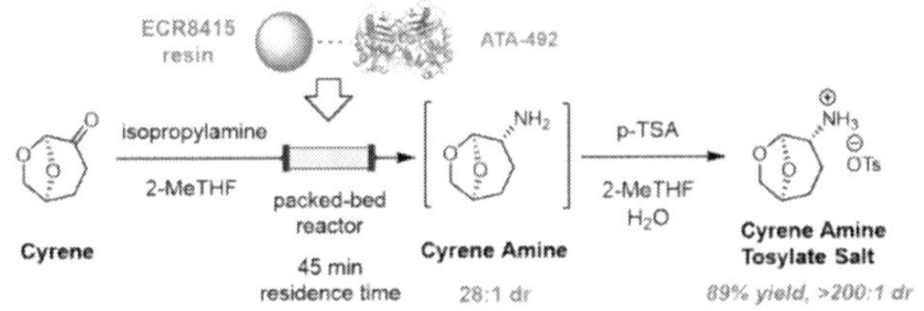

Final process reaction profile, performed at production facility, 10 L column

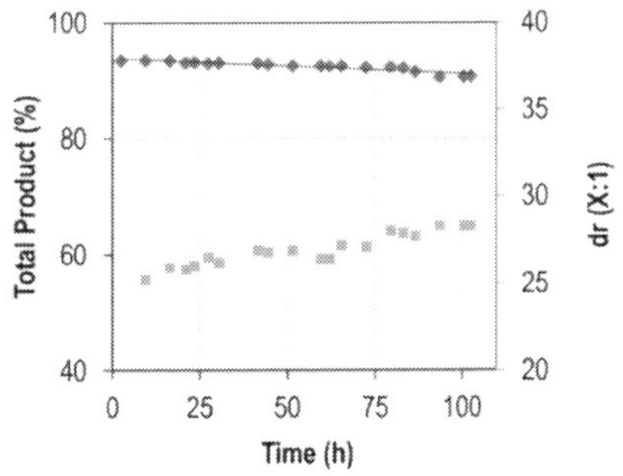

Process Changes significantly improved Transamination efficiency and sustainability

d.r. = 23:1.	→	d.r. >200:1.
2 kg ATA	- 55%	0.9 kg ATA
20 L column	- 50%	10 L column
3h RT 5 kg cyrene/day	- 75%	45 min RT 17 kg cyrene/day

Renewable Starting Material and Solvent

27

Transamination on Production Scale

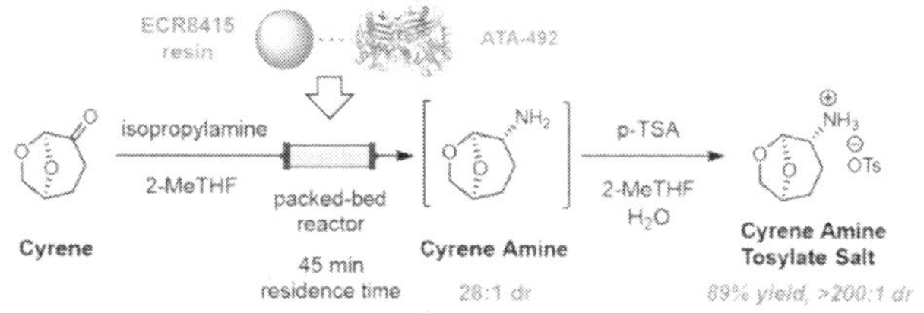

Final process reaction profile, performed at production facility, 10 L column

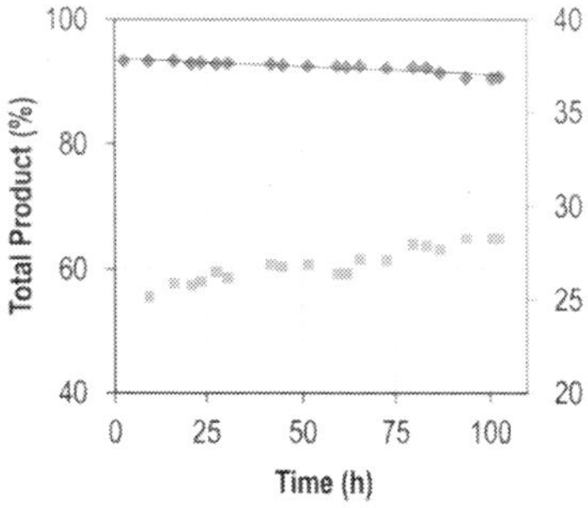

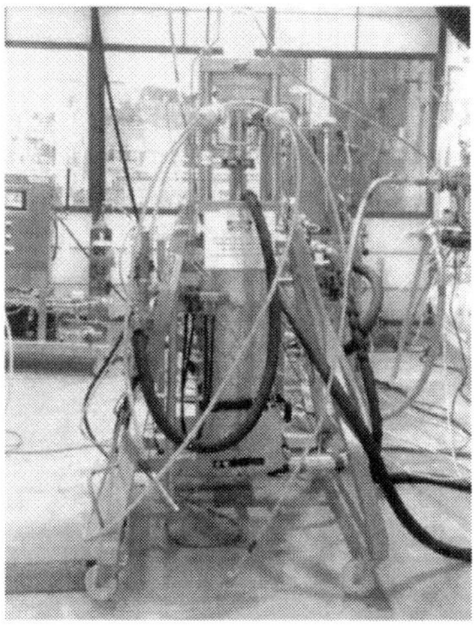

20 L DAC column used in pilot plant campaign

25

MK-1026 Commercial Route

Two-step synthesis of amino alcohol from cyrene

- Direct two-step route reduces time and resources required for manufacture
- Inputs are two bio-based solvents (cyrene and MeTHF) and isopropylamine
- Highly efficient evolved transaminase (used at 1.8 wt%)
- Immobilized enzyme system allows simple isolation from MeTHF
- Flow process streamlines reaction execution, reduces scale-up risks

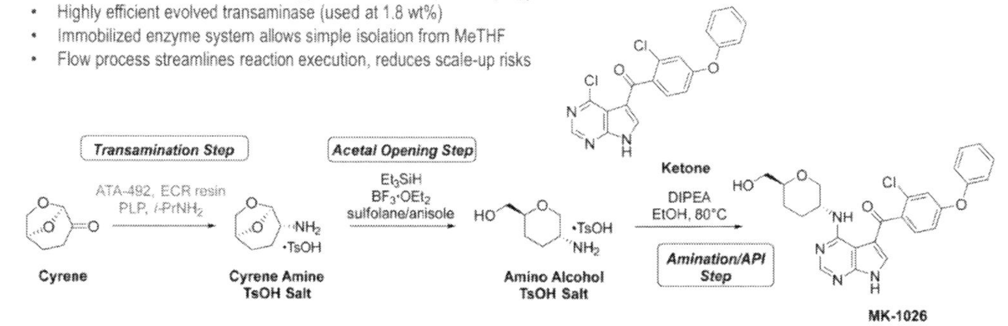

Renewable Feedstocks | Prevent Waste | Atom Economy | Reduced Time and Cost

Acknowledgements

MK-1026 Transaminase Team

Xiaodong Bu
Matthew Burris
Karla Camacho Soto
Wai Ling Cheung-Lee
Cheol Chung
James Corry
Richard Desmond
Mike Di Benedetto
Mike Di Maso
Lien Do
Jacob Forstater
Shane Grosser
Erik Guetschow
Joe Gouker
Hsing-I Ho
Erik Hoyt
Tetsuji Itoh
Lisa Jellet
Nadine Kuhl
Jeffrey Kuethe
Latevi Lawson
François Lévesque
Anne Maguire
Amanda Makarewicz
Kevin Maloney
John McIntosh
Jon McMullen
Jeffery Moore
Grant Murphy
Karthik Narsimhan
Andrew Owens
Weilan Pan
Chris Prier
Ji Qi
Nelo Rivera
Val Rodrigues
Rebecca Ruck
Gao Shang
Glenn Spencer
David Thaisrivongs
Deeptak Verma
Zhixun Wang
Brian Wyvratt
Daniel Zewge

Codexis

Pierre Brazeau
Stefanie Minor

Pharmaceutical peptide drug substance purification using Nanofiltration

AIChE Process Development Symposium 2022

Paridhi Agrawal
Advisor - Engineering

Kevin D. Seibert
Associate VP - Engineering

Eli Lilly & Co.
Synthetic Molecule Design and Development

Lilly

Membrane Filtration

Size-based separation of components of a stream

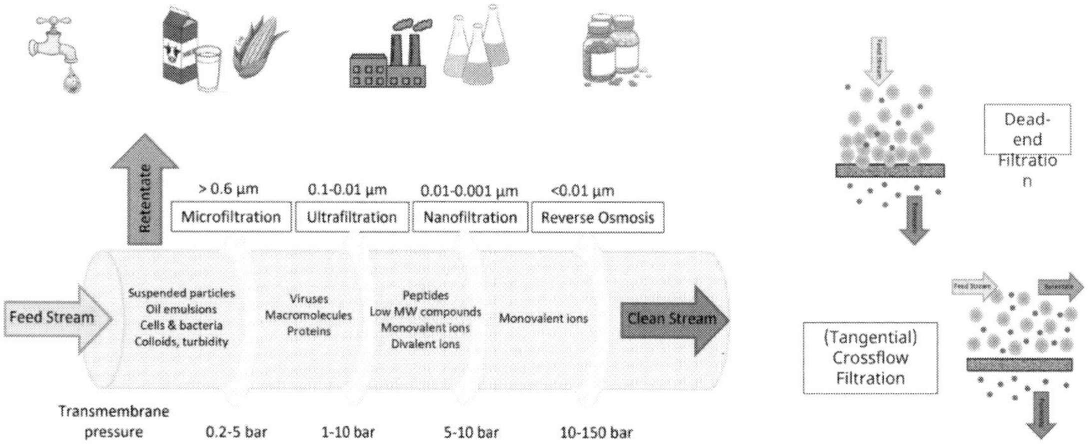

2

Applications in Pharmaceutical Industry

Small molecule API

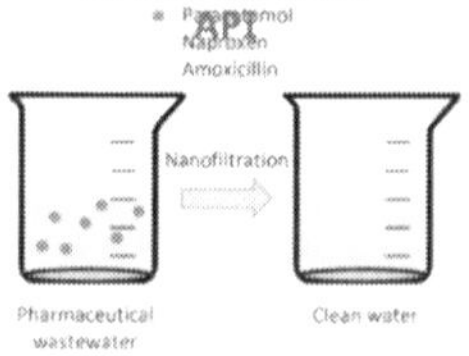

Injectable mAbs

Insulin

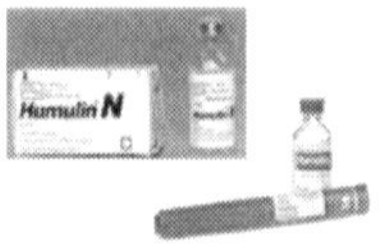

Future Directions for Pharma:

- Medium-large molecules as APIs (peptides and siRNAs)
- High production volume
- Continuous manufacturing

Precipitation of amorphous solids is generally challenging, especially on scale-up

- Poorly filtering and poorly washing solids
- Chemical and physical properties often make them prone to gelation
- Strong driver for eliminating isolations - through (continuous) processing

Need a unit operation to remove low MW impurities and avoid isolation of intermediates/product

Nayak et al. *Membranes 12.5 (2022): 528.*
Shahtalebi et al. *IJEHSE SPRING 8.2 (2011): 109-116*

Paradigm Shift in Peptide Synthesis

Market Challenge: Increasing importance of longer, complex peptides with high production demand

Recombinant Synthesis

- ✓ Efficient; high chiral purity
- × Natural occurring amino acids only
- × Long development timeline

Solid-phase Peptide Synthesis

- ✓ Gold standard for unnatural peptides since last 5 decades
- ✓ Automatable process
- × Complex impurity profiles
- × Poor yield for long fragments

Hybrid SPPS-LPPS

- ✓ Unnatural AAs for PK/PD advantages
- ✓ Complex sidechains for many peptides
- ✓ Route flexibility for coupling difficult peptide residues
- ✓ Campaign-able intermediates
- ✓ Decreased risk of late-synthesis failure
- ✓ High purity
- ✓ Fits traditional small-molecule facilities
- ✓ Amenable to continuous processing – automation + less material at risk

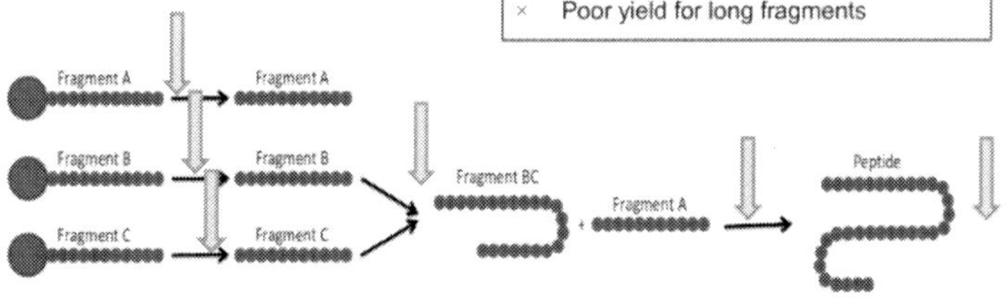

Nanofiltration in Hybrid Peptide Production

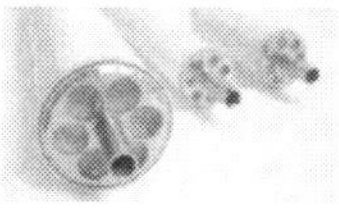

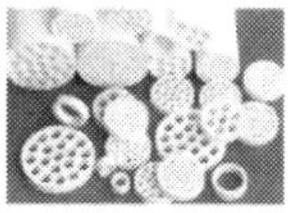

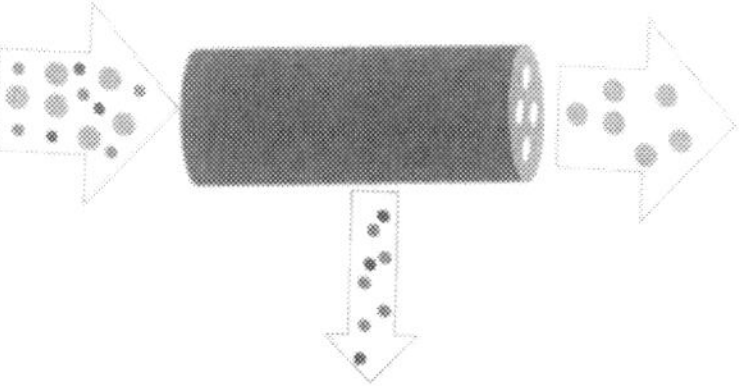

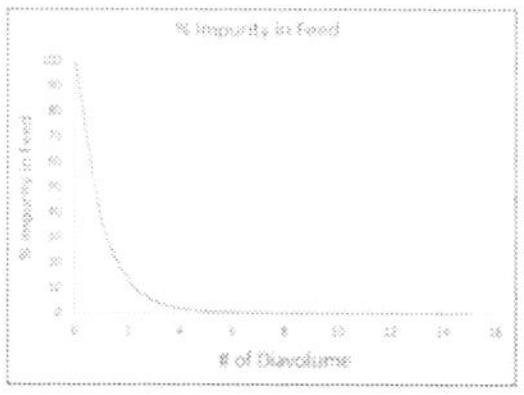

Coupling and synthesis		
Compound	**Role in synthesis**	**MW**
N,N'-diisopropylcarbodiimide (DIC)	Activator	126
Diisopropylurea	By-product of activation by DIC	144
HATU	Activator	379
Tetramethylurea	By-product of activation by HBTU	116
HOBt	Racemization suppressor	135
Diisopropylethylamine (DIPEA)	Base for facilitating activation	129
PyOxim	Activator	527
Deprotection		
Compound	**Role in synthesis**	**MW**
Piperidine	DeFMOC Reagent	85
Diethylamine (DEA)	DeFMOC Reagent	73
Dibenzofulvene	By-product of DeFMOC process	178
DBF-piperidine	By-product of DeFMOC process	263

Name	Type	MWCO
Polymeric	Duramem	150, 200, 300, 500, 900
	Puramem	Selective, Performance, Flux, S600
	AMS	100, 200, 400, 2500
Ceramic	Hydrophilic	200, 450, 600, 750, 2000, 3000, 5000, 8000
	Hydrophobic	450, 600, 750, 2000, 3500

- ✓ Physical separation based on molecular weight of species
- ✓ Eliminates challenging amorphous solid isolations
- ✓ Removes byproducts which impact downstream chemistry
- ✓ Diafiltration washes out impurities (by theoretical CSTR)
- ✓ Despite low temperature operation, thermodynamically unfavorable solvent exchange possible

5

Process Schematic: Dual-pump Nanofiltration

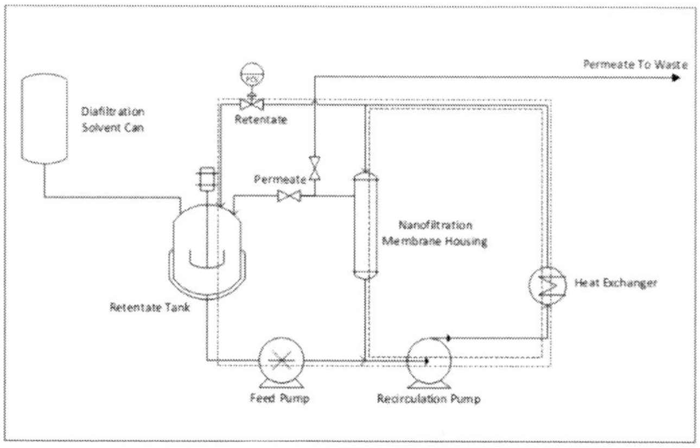

Process parameters:

- Feed flow rate
- Crossflow rate
- Transmembrane pressure
- Temperature
- Membrane material, dimensions, pore size
- Solution viscosity
- Peptide concentration and loading

6

Nanofiltration Fundamentals and Process Goals

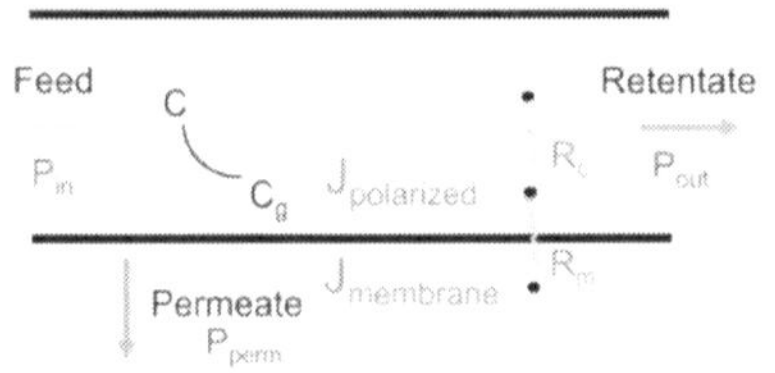

$$Q_{perm} = J_{resist} A_{mem} \qquad J_{resist} = \frac{TMP}{\mu(R_m + R_c)}$$

$$TMP = \frac{(P_{in} + P_{out})}{2} - P_{perm}$$

$$\frac{1}{J_{resist}} = \frac{1}{J_m} + \frac{1}{J_{polarized}} \qquad J_{polarized} = k \ln\left(\frac{Cg}{C}\right)$$

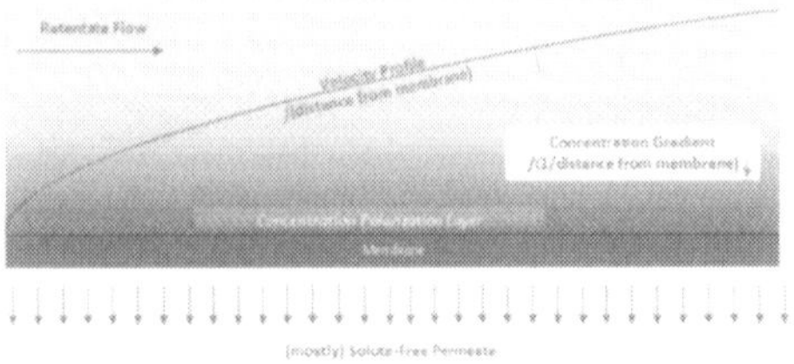

- Peptide accumulated on membrane provides additional resistance to solute permeation
- Measure development by drop in permeate concentration

Design Goals/Optimization Metrics:

- Yield (peptide retention)
- Cycle time (flux)
- Membrane life (fouling)

7

Modes of Filtration

Setup peptide layer on membrane for enhanced retention

Remove peptide from membrane before next cycle

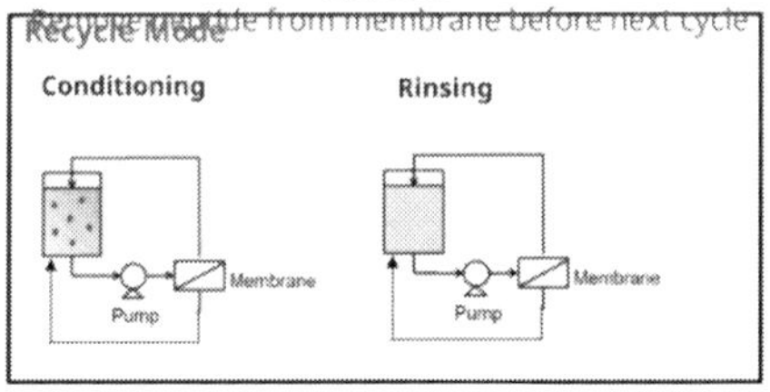

Concentrate feed by solvent removal

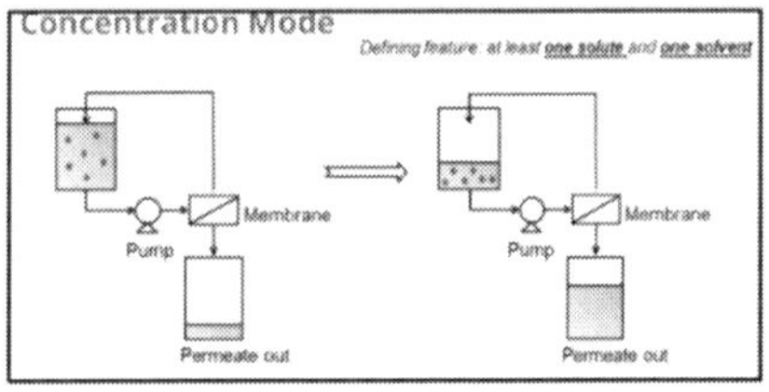

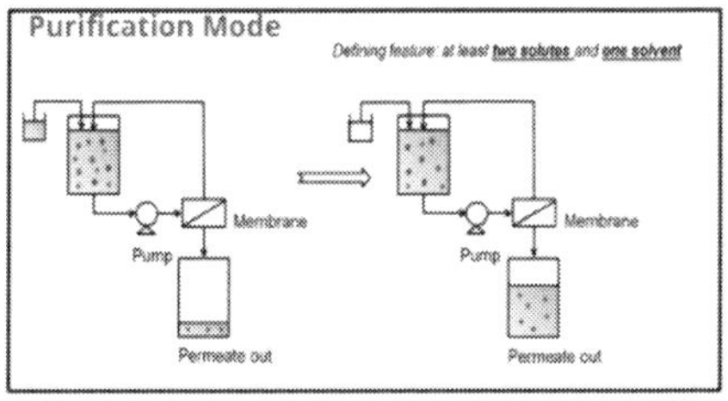

Diafiltration: Purify feed by low MW species removal

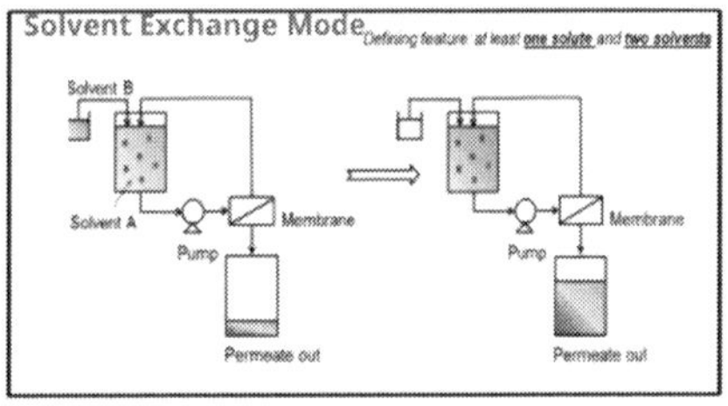

Diafiltration: Solvent swap

8

Case Study: Synthetic Peptide Purification

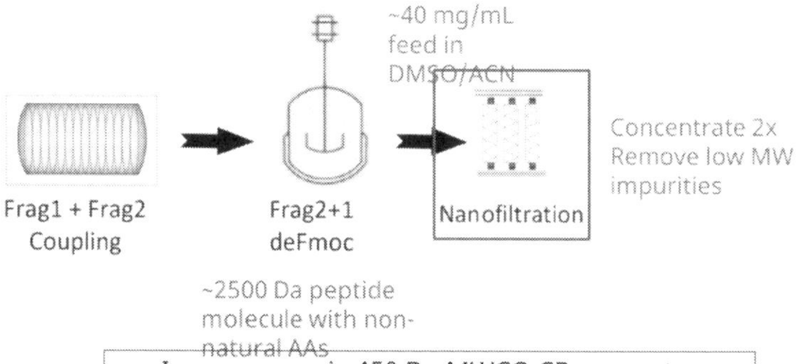

Inopor ceramic 450 Da MWCO CB geometry membrane
Up to 18 bar TMP and 6 m/s crossflow velocity tolerable
Peptide sequence agnostic

How do controllable operating parameters (crossflow rate and TMP) affect NF performance?

9

Pilot Scale Nanofiltration Equipment

Delta-V operated NF skid
Online HPLC for near real-time process monitoring

10

Peptide Retention Study

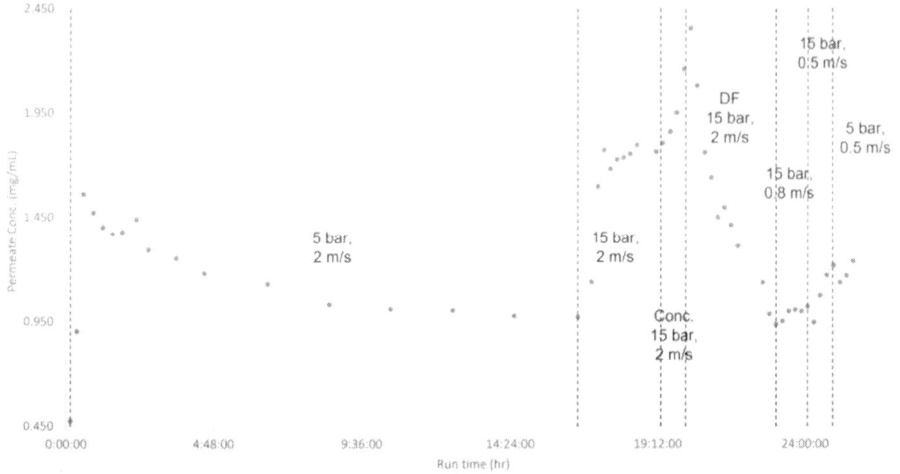

High crossflow rate and low TMP improve peptide retention

11

Flux Study

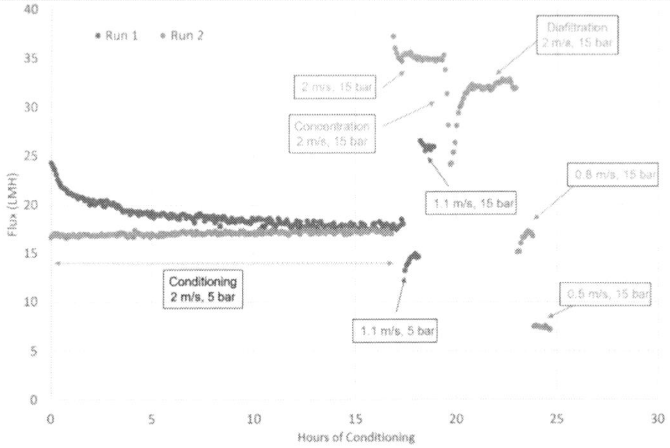

High crossflow rate gives high flux; high TMP gives high flux

High crossflow rate and low TMP reduce peptide buildup/membrane

12

Key Takeaways

- With the increase of larger API molecules in development pipelines, and need of high-volume manufacturing, continuous purification bypassing isolation needs to be developed
- Nanofiltration can be used for concentration, purification and solvent swap of synthetic peptides
- Pump crossflow rate and TMP are tunable engineering parameters in NF
- High crossflow rate and low TMP improve peptide retention and reduce fouling
- Crossflow rate and TMP are directly proportional to flux

Design Goals/Optimization Metrics:

- ✓ Yield (peptide retention)
- ✓ Cycle time (flux)
- ✓ Membrane life (fouling)

Other Considerations in NF Development:

- Manufacturing quality of membranes (assess by clean solvent flux and PEG retention)
- Hydrodynamic flow regime in membrane channels (turbulent flow ensures flat velocity profile & consistent scale-up)

13

Acknowledgement

Synthetic Molecule Design & Development
Engineering Tech Center
D&M Continuous Solutions
Lilly Kinsale Manufacturing
Inopor

- Kevin Seibert
- Nil Tandogan
- Nick Klitzing
- Frank Torres Torres
- Doug McKinney
- Bruce Parker
- Cyrus Fisher
- Katie Williams
- Kyle Powers
- Carolina Martinez
- Morgan Rosemeyer
- Mark Berglund
- Darragh McDonagh
- Muiris Macphaidin
- Sean Doherty
- John Howard
- Michael Reen
- Marie O'Mahony
- Aideen O'Malley
- Gearoid Finn
- Ashley Humenik
- Jinju James
- Joel Calvin
- Jennifer Groh
- Ric Miller
- Zach Harms
- Kristine Aldulescu
- Brad Campbell
- Raymond Boyse
- Carmel Condon
- Aoife Corrigan
- Declan Gavin
- Olivia Gowran
- Caoimhe Keogh Hansen
- Sharon Kelly
- Laura Maguire
- Nessa Mullane
- John Murphy
- Rosella Maria O'Mahony
- Richard Spencer

14

Questions?

TECHNOLOGY ADVANCES TO DRIVE WORKPLACE SAFETY

Kathy Kas, PE
Greg Nesmith, PE
Gary K. Johnson, PE
Dow Chemical

Process Safety
Risk Management

Dow
General Business

KATHY KAS, PE

- The Dow Chemical Company
- Manufacturing & Engineering Fellow, EH&S
- BChE - University of Delaware
- M.B.A. – Rutgers, The State University of NJ
- 37 years of industrial experience with Rohm and Haas/Dow
- 10 years with the Loss Prevention Symposium
- Book committees: *Guidelines for Initiating Events and Independent Protection Layers in Layer of Protection Analysis (LOPA), Guidelines for Investigating Chemical Process Incidents, Managing Cybersecurity: A Risk-Based Approach*
- Fellow of AIChE and CCPS
- CCPS Certified Process Safety Professional

Dow
General Business

Technology Advances to Drive Workplace Safety

AGENDA

- Case for Change
- History: Dow and Robotics
- Robotic Technology Applications
- Hazard Identification
- Implementation of a Corporate sUAS/Robotics Program
- Conclusions
- Q&A

Dow
General Business

Evolution of Industry

Industrial Phase	Century	Key Technologies
1.0	18th	Machines, steam and water-power, corporations
2.0	19th	Electricity, internal combustion engine, mass production
3.0	20th	Transistors, circuits, computing, software
4.0	21st	Internet of things, cyber systems, artificial intelligence

Dow
General Business

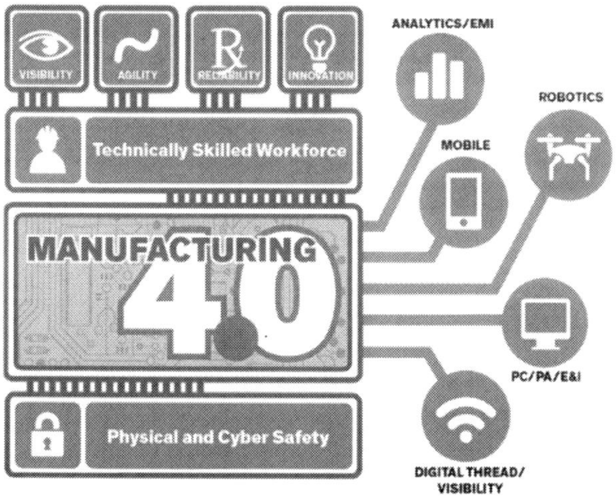

Case for Change

According to data collected by the U.S. Department of Labor…

- ~ 150 fatalities occurred during confined space entries in 2018
- > 800 workplace fatalities were attributed to falls to a lower level in 2020

Table 1. Fatal occupational injuries involving confined spaces, 2011-2018

Year	2011	2012	2013	2014	2015	2016	2017	2018
Fatal work injuries	120	88	112	116	136	144	166	148

Table 2. Fatal occupational injuries for selected events or exposures, 2016-20

Characteristic	2016	2017	2018	2019	2020
Fall, slip, trip	849	887	791	880	805
Fall on same level	134	151	154	146	136
Fall to lower level	697	713	615	711	645
Fall from collapsing structure or equipment	65	48	50	37	36
Fall through surface or existing opening	87	85	83	95	63

Break-through technologies are needed to achieve these goals.

History: Dow and UASs/Robotics

- 2014: Initial projects targeted use of Robots and sUAS
- 2015: sUAS technology program began
- 2016: Piloting of sUAS/robotic technology
- 2018: Full deployment of Dow sUAS program, completing
  - From 2018 through 2021: over 5000 flights,
    - Over 4000 for CSE
    - Over 1000 for elevated work or site mapping
- 2018-21: Full deployment of the Robotics program, completing:
  - From 2018 through 2021: over 6000 Confined Space Entries avoided,
    - Over 4000 by use of sUAS
    - Over 2000 by use of robots, probes, and cameras
- Savings of tens of millions of dollars in avoided:
  - Downtime/productivity losses
  - Inspection/maintenance costs
  - Scaffolding costs

Dow General Business

Robotic Technology Applications

- Photography and video
- Mapping
- Facility/structural inspections
- Insulation integrity/process performance reflective of heat generation
- Sampling
- Non-destructive testing
- High-pressure cleaning
- Cutting and drilling
- Security/surveillance
- Emergency response
- Search and rescue
- Construction monitoring
- Accident investigations

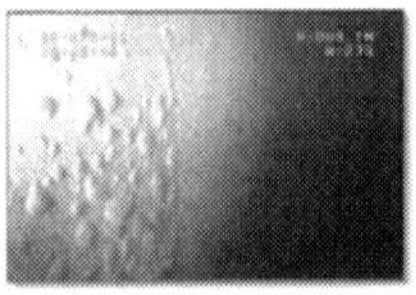

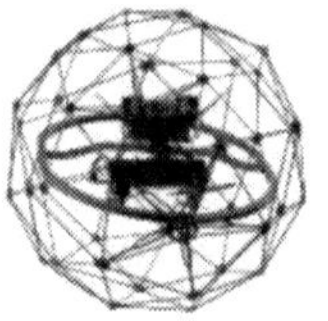

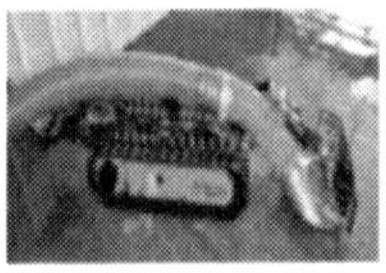

Dow General Business

Exciting tech! Let's do this!

Photo of drone with dead battery inside process vessel sitting in a pool of potentially damaging residual material.

What Went Wrong?

Error	Root Cause
Dead battery	Flight path did not account for internal obstructions; poor planning
Hazardous Area Classification (HAC) incorrect	Misunderstanding of internal HAC; lack of expert review; 2nd drone also an error
Drone materials of construction incompatible	No reactive chemicals review
Gas detection at entrance only	Emphasis on people exposure and not equipment compatibility
Static electric discharge from propellers not considered	No formal hazard identification or risk assessment

Dow General Business

Let's try this again...

The Submarine Robot (Mini-Sub)

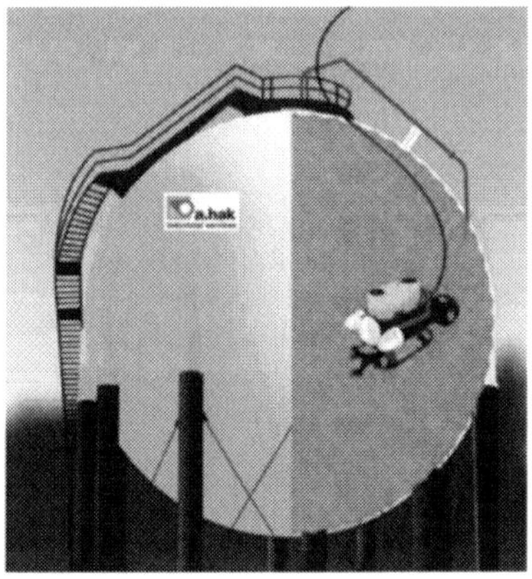

No Confined Space Entry!

Dow General Business

Mini-Sub Hazard Identification

Issue	Response/Result
Flammability	Propylene removed; filled with water
HAC and Ignition Sources	Filled with water; no oxygen for ignition
Human exposure to chemicals	Decontamination of mini-sub upon removal reviewed and approved
Adding/removing robot	Safe procedure developed and approved
Other issues	Consulted with vendor; approved
Reactive chemicals	Robot parts are compatible

Then finally, someone asked...

Use a multi-discipline team for your reviews!

Use of sUAS and Robots - Basic Considerations

Pre-planning of robotic operations is critical:

- What are the objectives of the operation?
- What device(s) will be needed?
- Who will be conducting the activities?
- Where will the activity occur – and what hazards might exist or be created?
- What processing operations might occur during the same time as the robotic activity?

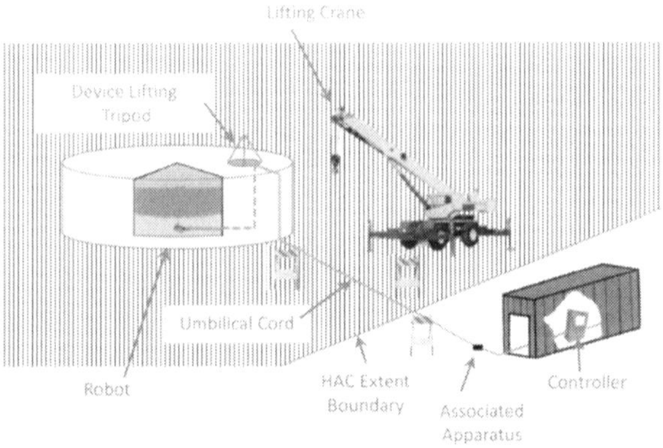

DOW General Business

Use of sUAS and Robots - Basic Considerations

- Pilot certification and sUAS registration
- Data transmission frequency
- Pilot/operator location and line-of-sight
- Location of battery/charging unit
- Equipment inspection
- Communication protocol (pilot/operator, visual observer, plant personnel)
- Cessation of process/maintenance work during activity
- Conditions when robotic activity needs to be suspended

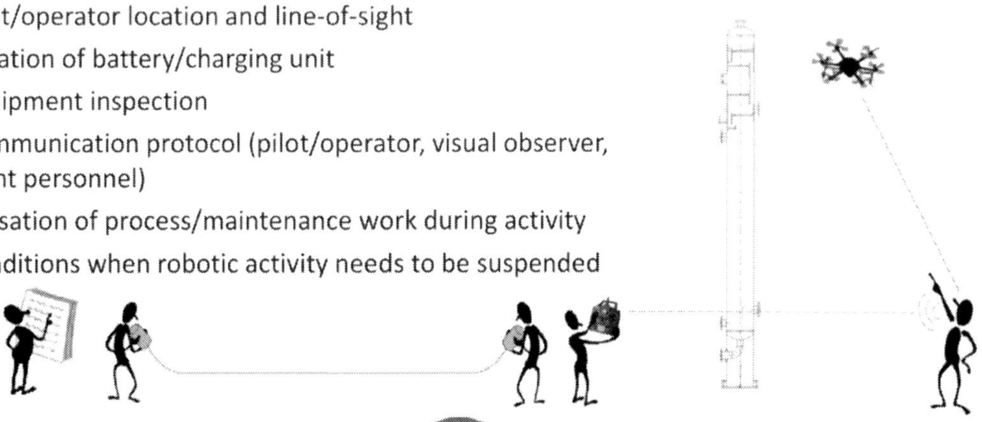

DOW General Business

Use of sUAS and Robots - Travel Path Considerations

- No-fly zones or private land
- Elevated structures
- Flight instability zones
- Nesting birds
- Collision with critical equipment
- Snagging of robotic cables
- Loss of power
  - ➢ Launch/battery charge locations
  - ➢ If the device fails, where will it land?
- *Potential condition changes over time*

Plan travel in four dimensions!

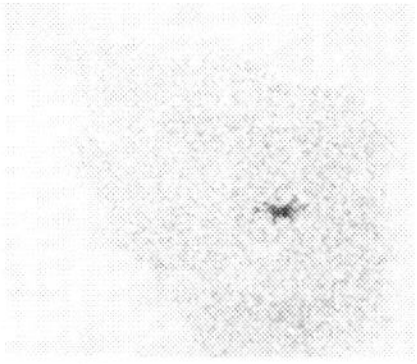

Use of sUAS and Robots *External* to Process Equipment

Will the device pass through/over, or conduct operations in, a classified area?

- Sumps
- Trenches
- Tank breather vents
- Pressure relief devices
- Process vents
- Open floating roof tanks

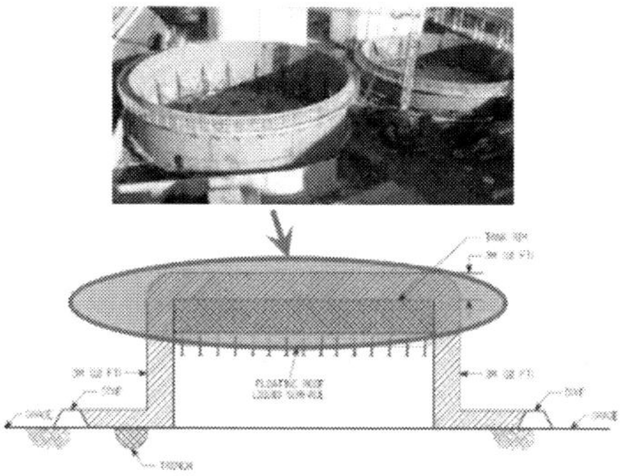

Three options for working in classified areas:

- Select a system that is rated for the HAC of the travel route <u>and</u> the inspection area
- Remove materials to temporarily change the HAC rating
- Utilize a Hot Work Permit and safe operating discipline

Best Practice – Draw the travel path on a copy of the Hazardous Area Classification Drawing (HACD), indicating flight elevations and hazards to avoid.

Use of sUAS and Robots – *Is the Device Properly Rated?*

Caution:

- Some suppliers have incorrectly represented their equipment as suitable for hazardous (classified) areas.
- Use of individual, certified components does not make a device certified.

> *Real Example:*
> A vendor indicated that a robot was certified for Class I, Division 2 service because the camera component mounted on the robot was certified. However, the assembled robotic unit was never certified as a system - and was actually not suitable for use in a hazardous (classified) location since the wiring methodology and power management methods were not compatible with the HAC service.

- US: Nationally Recognized Testing Laboratory (NRTL) certification
- ATEX: Notified Body (NOBO) certification for Zone 0/1 or Zone 20/21*
- IECEx: Notified Body (NOBO) certification for Zone 0/1/2 or Zone 20/21/22

*Declaration of Conformity may be acceptable for Zone 2/22 applications with SME support.

Dow
General Business

Use of sUAS and Robots *Inside* Process Equipment

Internal operations present unique challenges, with the potential for:

- Injury/exposure to personnel introducing/removing the device.
- Residual flammable, toxic, pyrophoric, water/oxygen-reactive materials, especially in low spots, column packing, and hard-to-clean areas
- Damaging process equipment.
- Contaminating process equipment.

External sUAS

Internal sUAS

Dow
General Business

Internal Use – "Prop Wash"

"Prop wash" (air movement from sUAS rotors) can present special hazards:

- Prop wash (and hydroblasting, vacuuming, etc.) could release trapped ignitable material from residual sludge or create an ignitable dust cloud from combustible dust layers.
- Particulate movement or atomization of residual material from prop wash can increase the internal concentration of flammable materials.
- Prop wash can create static electricity; be cautious if the equipment contains low Minimum Ignition Energy (MIE) materials.
- Prop wash can create combustible dust clouds.

Dow had not yet evaluated commercial sUAS for use in combustible dust applications.

Dow
General Business

Use of sUAS and Robots *Inside* Process Equipment

Other important safety considerations for internal operation:

- Emptying and cleaning doesn't change the surrounding HAC.
- Ventilation inside equipment may create flight instability.
- Hot internal surfaces; certifications have temperature limits
- Lack of line-of-sight; challenging flight path
- Equipment failure inside of the process equipment

> ***Real Example:***
> A sUAS was in use inside a process vessel. However, it was difficult for the pilot to maneuver around obstacles, and the operation required more flight time than expected. When the low power alarm was received, the pilot immediately began the extrication flight. Unfortunately, because of the complicated return flight path, the sUAS ran out of power before it could reach the manway, and it fell into the bottom of the vessel. Retrieval was later successful by "hooking" the failed UAV with a "rescue" sUAS.

Dow
General Business

Robotics *Inside* Process Equipment WITH Process Chemicals

Potential issues:

- Area below tank level may be unclassified, but device may pass through classified headspace. Manage carefully.
- Inerting may create other process hazards or asphyxiation risks.
- Device may contaminate the process chemical.
- Process chemical might damage the device.
- Decontamination plan is required - or the device might be disposable.

> *Real Example:*
> A plant was evaluating the use of a robot to inspect a tank without removing the process chemical. The metallic components were evaluated and found to be compatible with the process chemical. However, it was subsequently determined that the soft, elastomeric components of the robot would have dissolved while immersed in the process material.

Implementation of a Corporate Robotics Program

- Corporate Standard
  - General policies and guidance
  - Roles and responsibilities
  - Legal requirements
  - Limitations and boundaries for use/when expert review/approval is required
  - Expert subject matter contacts
  - Management of Change
- Approved Device/Contractor List
  - Pre-approved robotic devices (and for what applications they are approved)
  - Pre-approved contractors (by region)
- Training
  - By role (e.g., general awareness, approver, operator)
  - Multiple languages
- Communicating "Success Stories" and "Lessons Learned"

Implementation of a Corporate Robotics Program

- Checklist(s)
  - Hazard identification/mitigation
  - Documented flight path on HACD; include hazards to be avoided
  - Permit requirements
  - PPE requirements
  - Conditions requiring cessation of work
  - Emergency retrieval plan
  - Verification that device and pilot are properly registered/certified
  - Pre-operation equipment inspection
  - Communication plan (pilot/operator, visual observer, plant personnel)
  - Documentation of review and approval by key individuals

Recommendations

- Intentionally search for hazards BEFORE employing any new technology in a processing area.
- Use organized, rigorous hazard identification tools and multi-disciplinary teams to prevent incidents before they occur.
- Use corporate standards, checklists, approved vendor/equipment lists, training, and a communications plan to execute new tech safely.
- History may be limited but learn from mistakes. Investigate and leverage learnings.

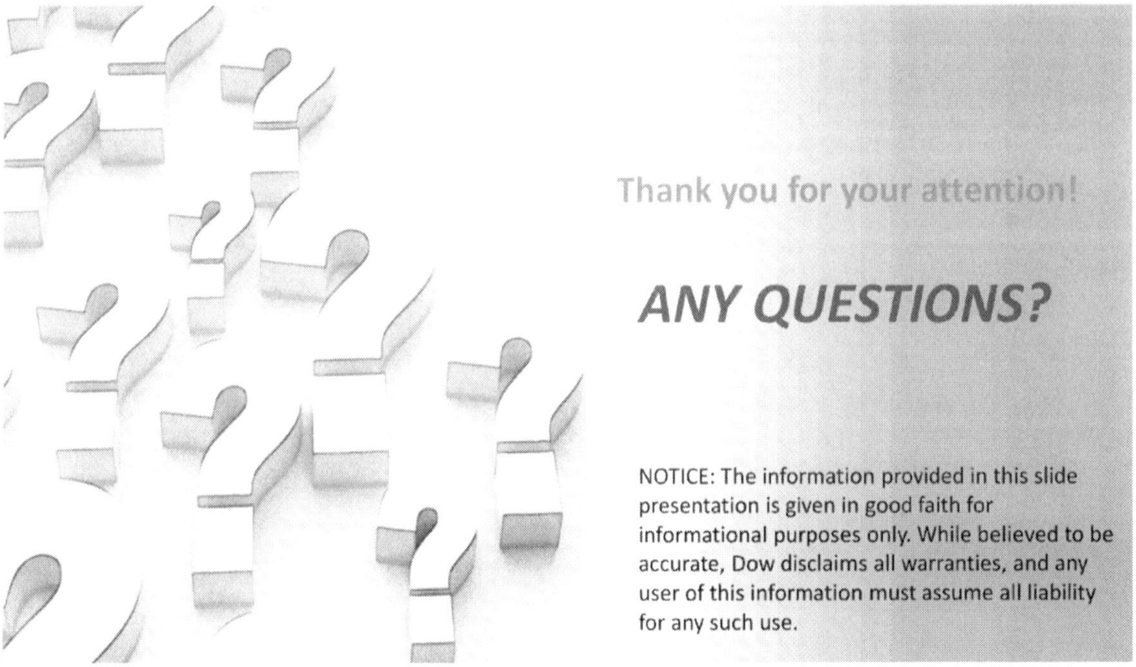
Thank you for your attention!
ANY QUESTIONS?
NOTICE: The information provided in this slide presentation is given in good faith for informational purposes only. While believed to be accurate, Dow disclaims all warranties, and any user of this information must assume all liability for any such use.

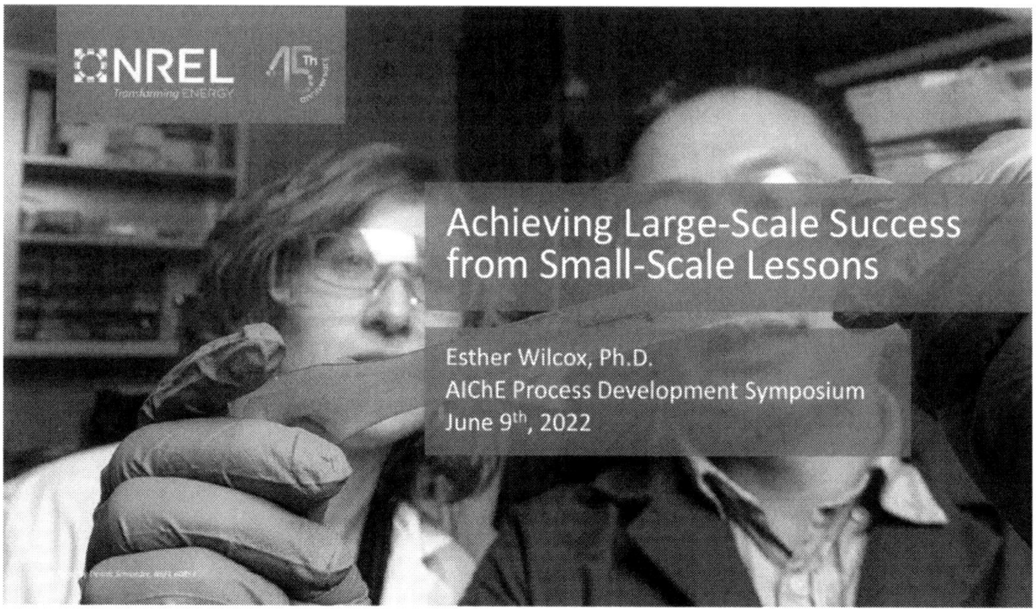

Diversity And Inclusion = Key To Success

Engineers, Operators, Researchers, Safety, Scientists, Technicians:

- Each one has a unique view and experience.
- No single group or person is more important than another.

NREL | 2

Journal Articles Include Positive Outcomes, Rarely The Valuable Hard Learned Lessons

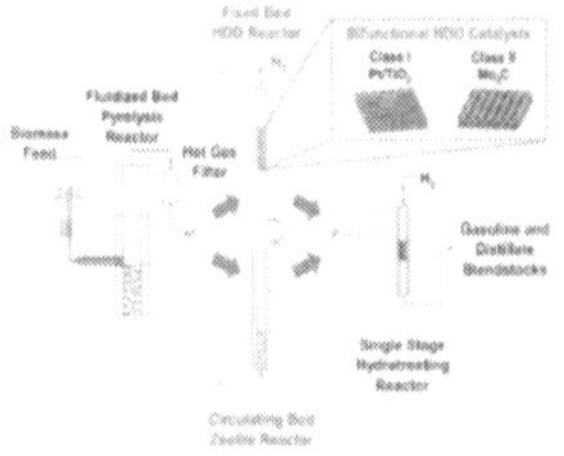

Energy Environ. Sci.,2018,11, 2904

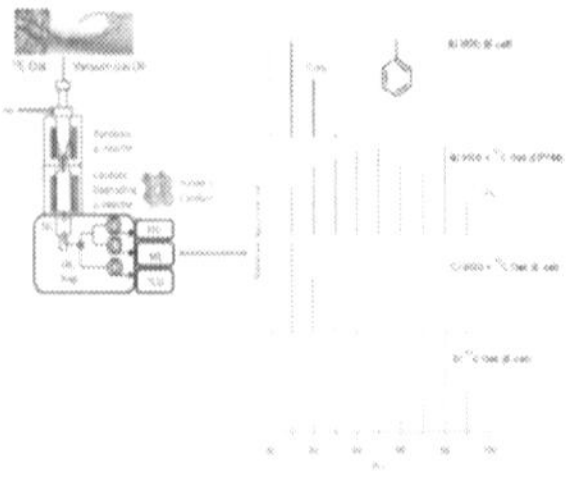

ACS Sustainable Chem. Eng. 2020, 8, 7, 2652–2664

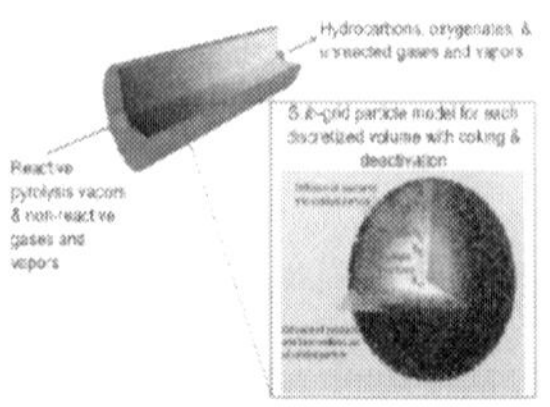

React. Chem. Eng.,2021,6,125

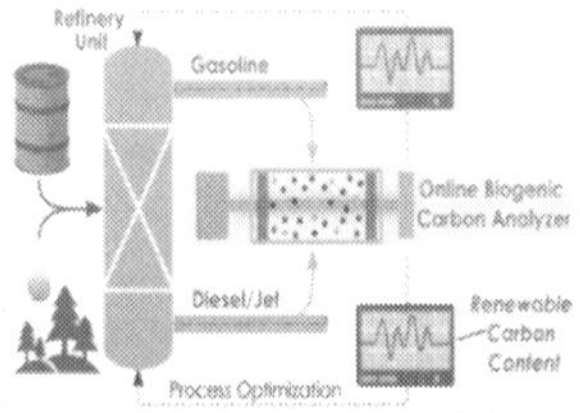

Anal. Chem. 2021, 93, 10, 4351–4360

Fundamental Science & Bench Scale More Than Just Exciting Data

A Valuable Learning Opportunity

- What did it take to achieve success?
- What ideas do you have for improvements?
- **What advice would you give for designing a pilot or full-scale plant?**

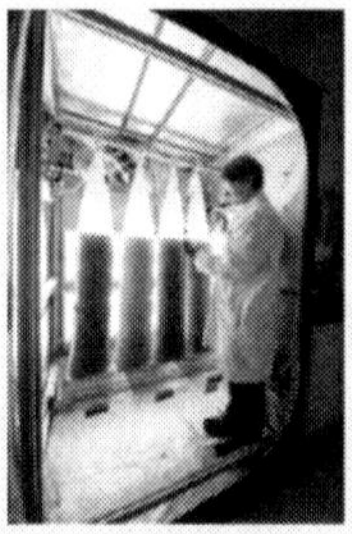

Photos by Dennis Schroeder / NREL

Safety Data Sheets Don't Always Tell The Whole Story:

Material Properties May Change During Or After Use

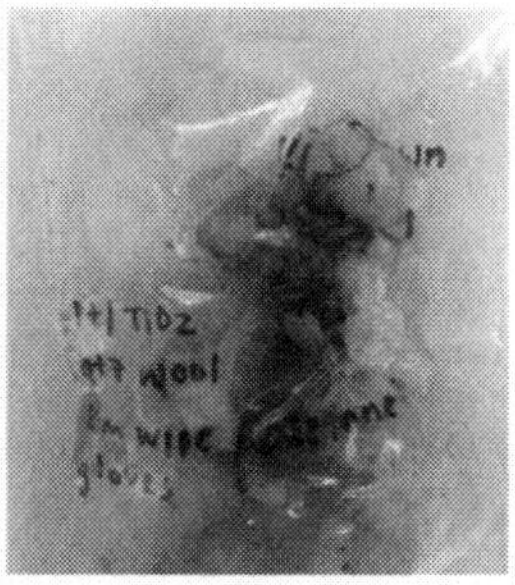

Scorched Bag in FTLB 251 (E-00520) 1/29/2019 – Photo by Dennis Smith / NREL

Debris from Hood Fire in FTLB 250 (E-00566) 5/10/2019 – Photo by Dennis Smith / NREL

NREL | 5

Making Safety Improvements Can Sometimes Have Unintended Safety Consequences

- Follow Management of Change Process even for safety improvements
- Include operators and technicians in change reviews

Example of gas cylinder automatic switchover – Photo by Esther Wilcox / NREL

Changed design for improved pressure safety – Photo by Christopher Golubieski / NREL

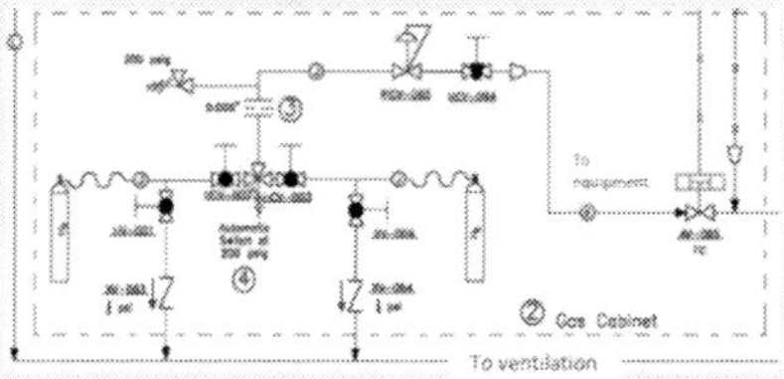

NREL | 6

Investigate Strange Things That Happen At Small-Scale To Prevent Incidents At Larger Scale

Debris from Thermochemical Process Development Unit (TCPDU) drum explosion Photo by Troy McCuskey / NREL

NREL | 7

Pilot Scale Is A Valuable Opportunity To Learn And Improve Safety And Operations For Full Scale

- What went well during the run?
- What did you observe that was not standard/normal?
- What issues were encountered?
- How did you resolve the issues?
- What ideas do you have for improvements?
- **What advice would you give for designing the next plant?**

Photos by Dennis Schroeder / NREL

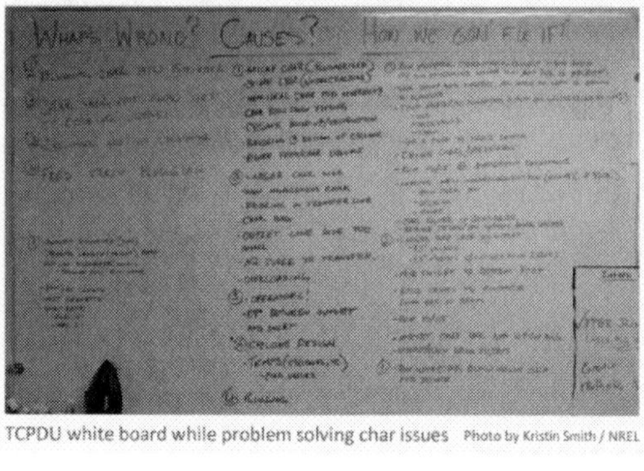

TCPDU white board while problem solving char issues Photo by Kristin Smith / NREL

NREL | 8

Seek Advice From And Listen To Technicians

- What are you observing during routine maintenance? Anything off-normal?
- What non-routine maintenance has been needed? How often?
- What ideas do you have for improvements?
- **What advice would you give for designing the next plant?**

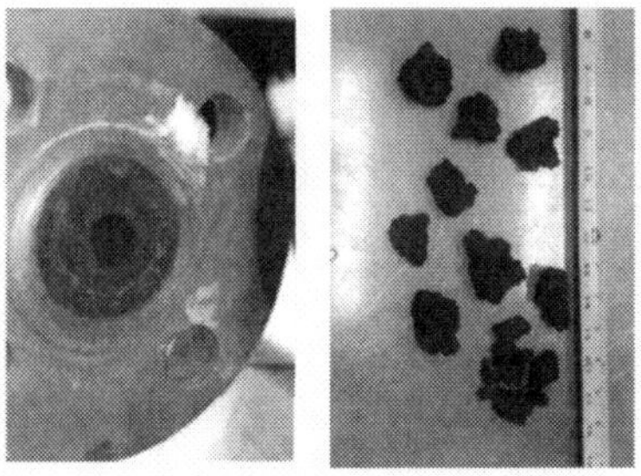

TCPDU plugged pipes and char in strainer

Photos by Christopher Golubieski / NREL

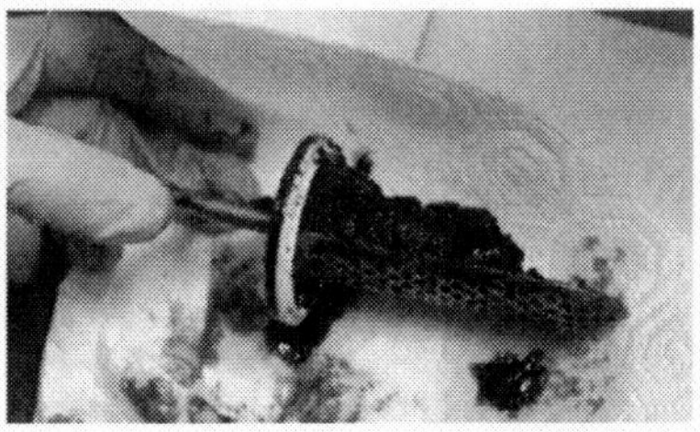

Technicians performing maintenance and repairs

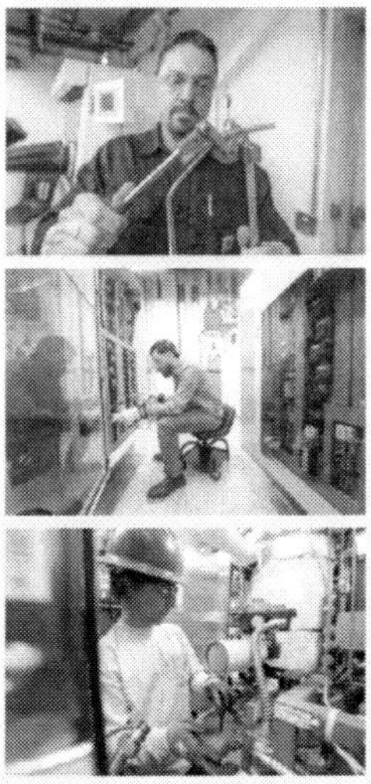

Photos by Dennis Schroeder / NREL

NREL | 9

Design With Maintenance & Repair In Mind

- Reduce repetitive motion and ergonomic issues
- Provide easy access where routine maintenance occurs

Technician must bend backward to look into pipe during routine maintenance

Photo from TCPDU camera / NREL

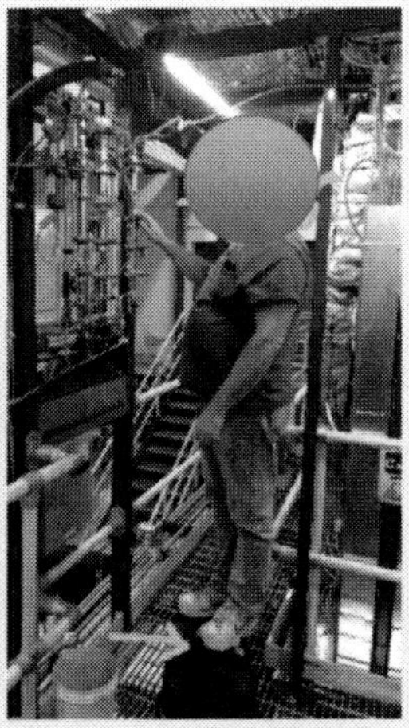

Technician must stand on stool and awkward reach to open/close valve

Photo by Lindsay Wilcock / NREL

NREL | 10

Everyone Wants The Technology Succeed

Researchers, Scientists, Operators, Technicians Are A Resource

Photos by Dennis Schroeder / NREL

NREL | 11

Phycological Safety Is Necessary For People To Be Willing To Share Ideas and Information

Create an environment where people feel safe to share.

Without blame or reprisal.

Where people feel heard and valued.

NREL | 12

Diversity And Inclusion
=
Key To Success

Involve Your Technicians, Operators, Researchers, Scientists

- Diversity includes experience, knowledge, and role.
- Inclusion means everyone can share their viewpoint, thoughts, and ideas safely.

Photos by Dennis Schroeder / NREL

NREL | 13

Product Development / Chemical Process Development

Mechanistic understanding guides impurity control strategy for CELMoD API reaction

AIChE Process Development Symposium
9-June-2022

Geoffrey Purdum, Jennifer Lott, Antonio Ramirez, Scott Bader, Antonio Ferretti, Hon-Wah Man, Richard Heid

Bristol Myers Squibb

A brief history of protein degraders @ BMS

- IMiDs (immunomodulatory imide drugs)
  - Modulate the immune system and other biological targets through multiple mechanisms of action

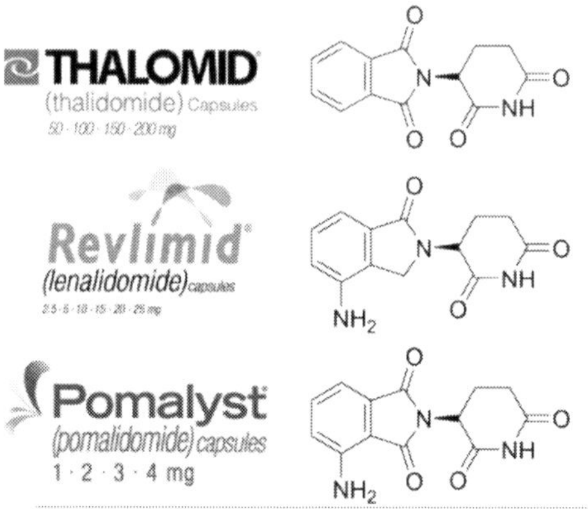

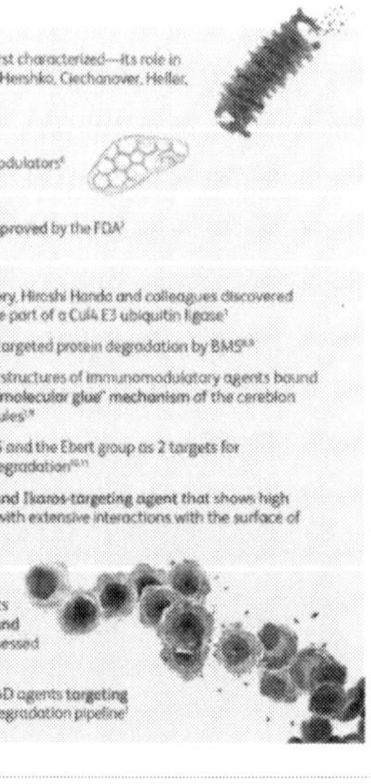

1980s
- The ubiquitin proteasome pathway is first characterized—its role in protein degradation is identified by the Hershko, Ciechanover, Heller, and Rechnsteiner groups

1990s
- BMS begins development of immunomodulators

2000s
- The first BMS immunomodulator is approved by the FDA

2010s
- In 2010, unraveling a 50-year-old mystery, Hiroshi Handa and colleagues discovered cerebion and proposed this protein to be part of a Cul4 E3 ubiquitin ligase
- **Cerebion** is identified as a mediator of targeted protein degradation by BMS
- In 2014, BMS published the first crystal structures of immunomodulatory agents bound to the cerebion complex, revealing **the "molecular glue" mechanism** of the cerebion surface and immunomodulatory molecules
- **Aiolos and Ikaros** are identified by BMS and the Ebert group as 2 targets for immunomodulator-mediated protein degradation
- **BMS began to investigate an Aiolos- and Ikaros-targeting agent** that shows high affinity with cerebion, which correlates with extensive interactions with the surface of cerebion

Present
- BMS has designed novel CELMoD agents with **observable immunostimulatory and tumoricidal** activity, currently being assessed across **multiple disease settings**
- There are several investigational CELMoD agents **targeting different proteins** in the BMS protein degradation pipeline

A brief history of protein degraders @ BMS

- IMiDs (immunomodulatory imide drugs)
  - Modulate the immune system and other biological targets through multiple mechanisms of action

THALOMID (thalidomide) Capsules 50 · 100 · 150 · 200 mg

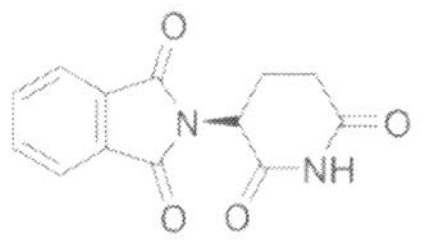

Revlimid (lenalidomide) capsules 2.5 · 5 · 10 · 15 · 20 · 25 mg

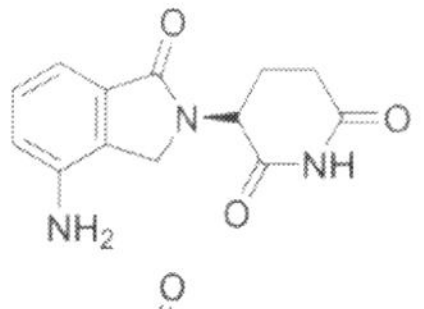

Pomalyst (pomalidomide) capsules 1 · 2 · 3 · 4 mg

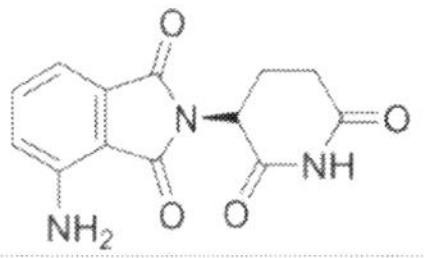

- CELMoDs (cereblon E3 ligase modulators)
  - Synthetic analogues of IMiDs
  - Act as a molecular glue binding to cereblon (CRBN) and recruiting proteins to the complex for eventual degradation

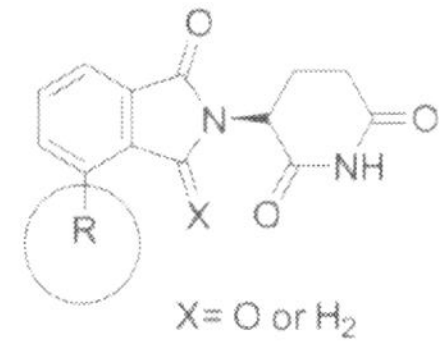

X = O or H_2

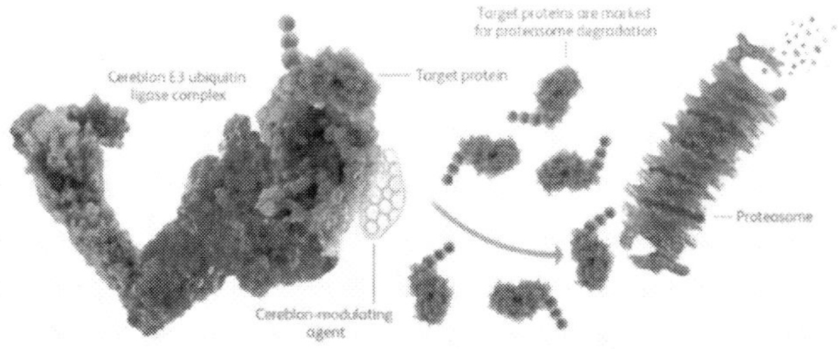

Safety considerations

- CELMoDs and IMiDs are highly-potent active pharmaceutical ingredients (HPAPI)
  - Known reproductive hazards
    - Thalidomide originally an OTC medicine in Germany in the late 1950s, resulting in thousands of birth defects and infant deaths
- EHS classifies this compound as ECB 5
  - OEL < 1 µg/m3 guideline, requiring significant engineering controls and committment to safety for process development and manufacturing
  - Limited up to 25 g per VBE for development work
  - In general, limited in-house and/or external HPAPI manufacturing capabilities

SAFC Verona facility for HPAPI manufacturing

https://www.pharmaceutical-technology.com/projects/safc-verona-hpapi-manufacturing-facility/

BMS Strategy for CELMoDs

- Glutarimide is responsible for CRBN binding, ultimately transforming the molecule into ECB 5
  - Advantageous to keep glutaramide formation as the final chemical transformation in the DS process
  - Two-step acid-catalyzed transformation developed to install glutaramide as final step

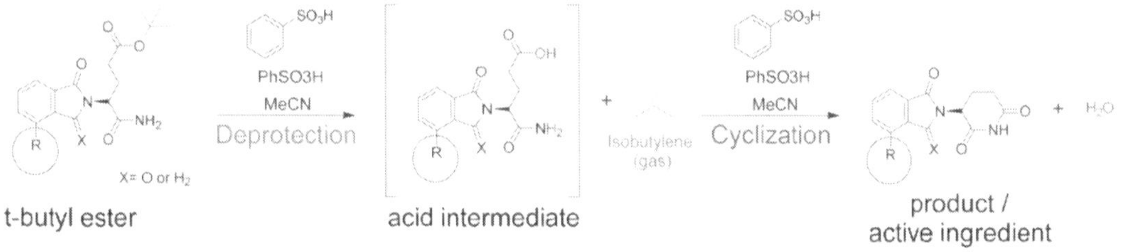

- High-potent nature of CELMoDs leads to low-commercial volume (<10 kg per batch)
  - Need robust process to minimize operational complexity while maintaining consistent batch-to-batch performance
- Goal: gain mechanistic insight into "universal" CELMoD API rxn to increase reaction robustness

CELMoD API rxn: Methodology

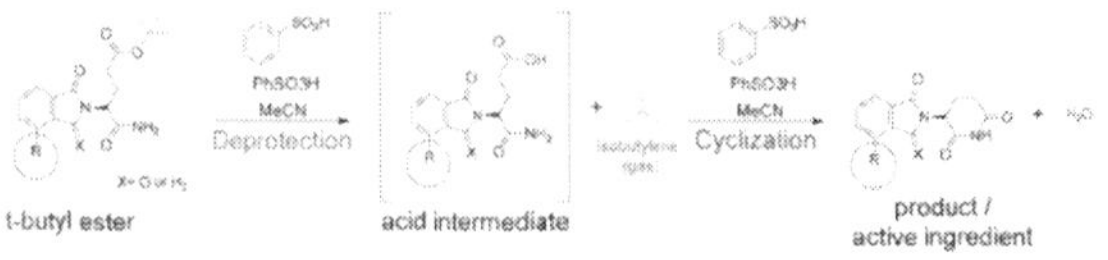

- Utilize in-situ measurements to gain insight into the reaction kinetics

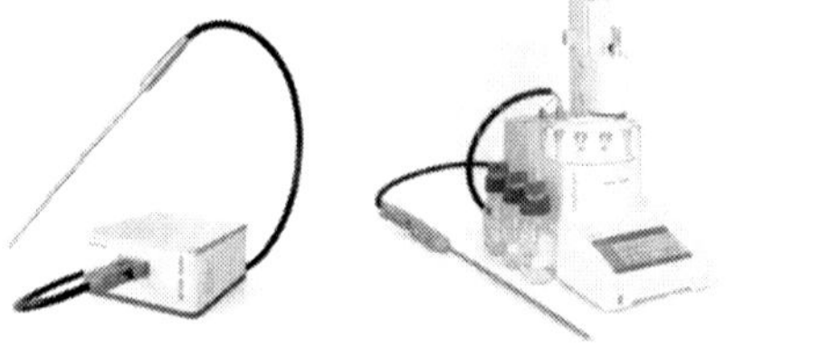

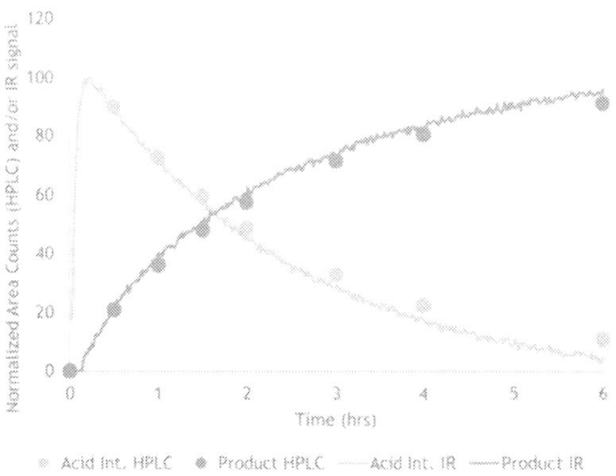

CELMoD API rxn: Methodology

- Utilize in-situ measurements to gain insight into the reaction kinetics

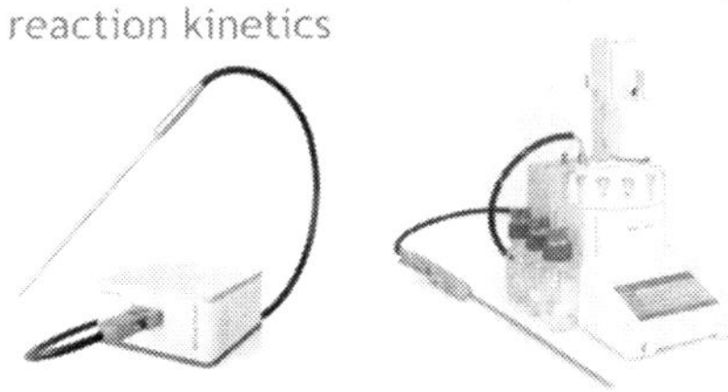

- Utilize model systems to decouple two-step reaction for mechanistic insight

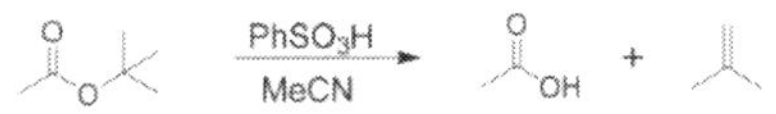

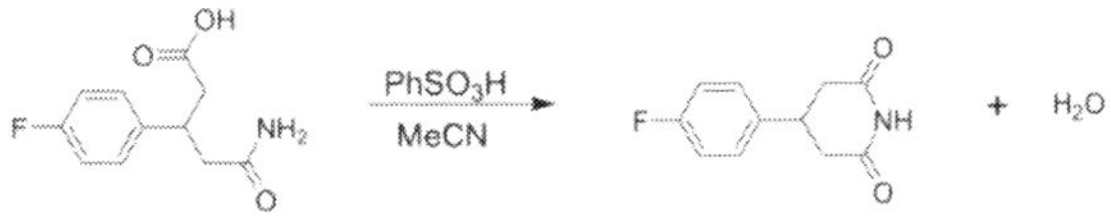

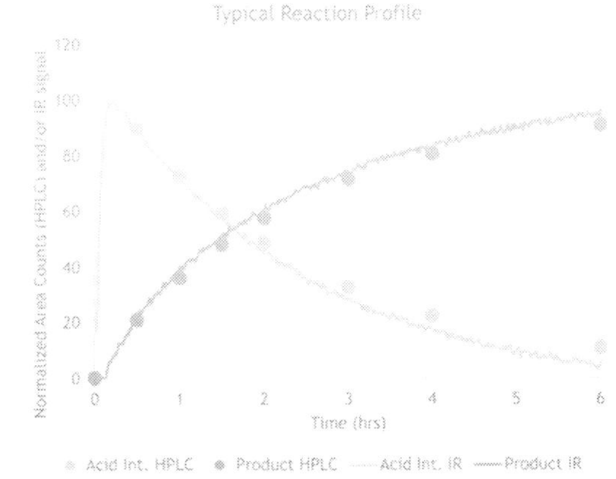

Desired reactions

Series of control experiments monitored via FTIR and HPLC to determine mechanism and fate of species

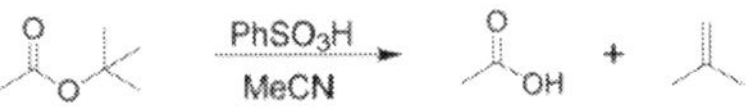

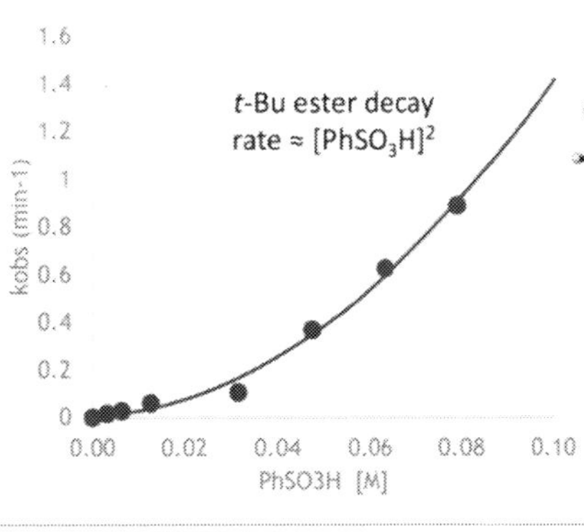

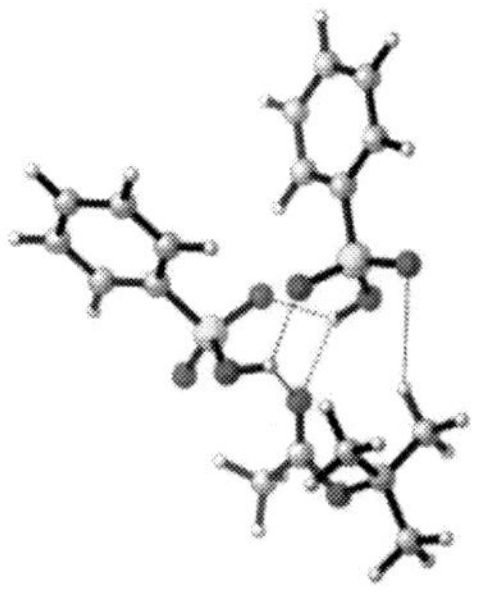

Desired reactions

Series of control experiments monitored via FTIR and HPLC to determine mechanism and fate of species

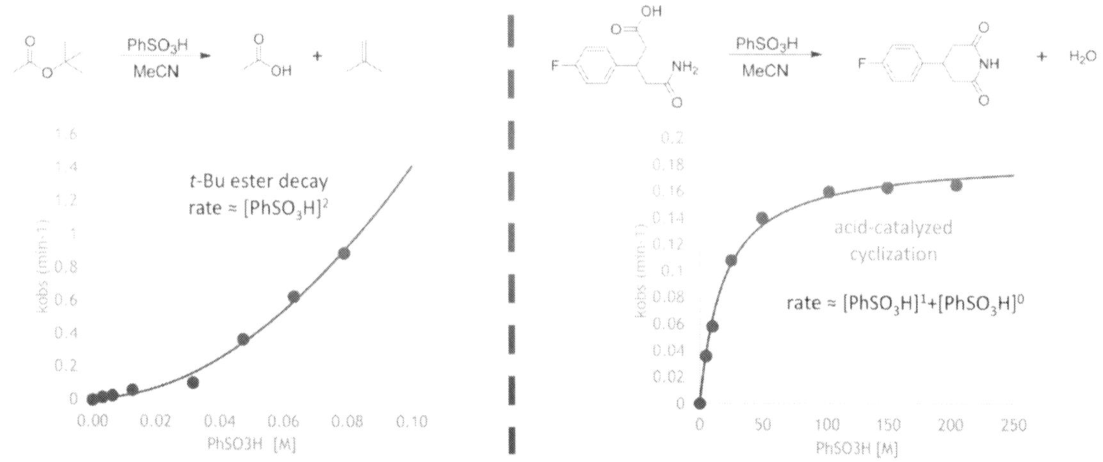

Bristol Myers Squibb | Product Development / Chemical Process Development

What about undesired reactions?

Many common impurities across the CELMoD assets

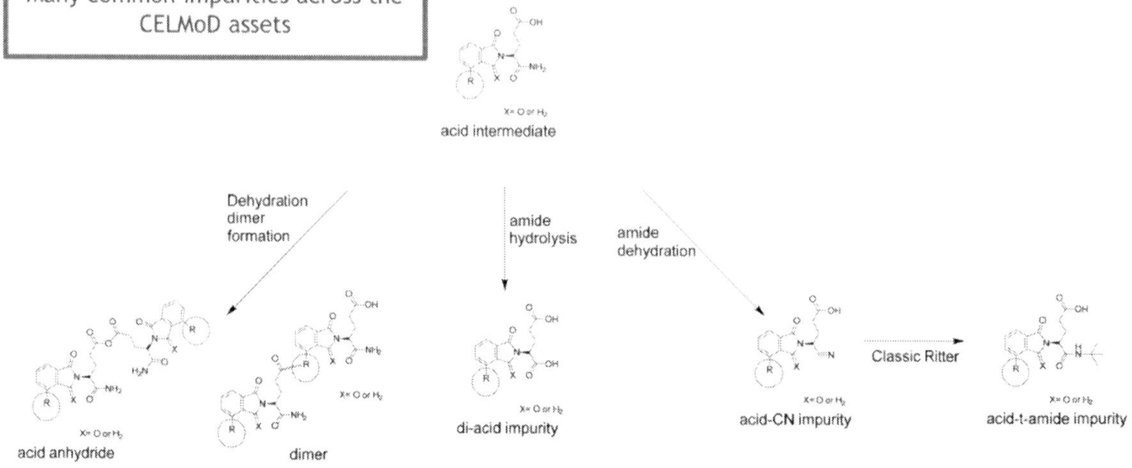

Bristol Myers Squibb | Product Development / Chemical Process Development

Proposed reaction pathways

10

What about undesired reactions?

Many common impurities across the CELMoD assets

Can we use model systems to understand reaction pathways?

What is the fate of isobutylene? Water?

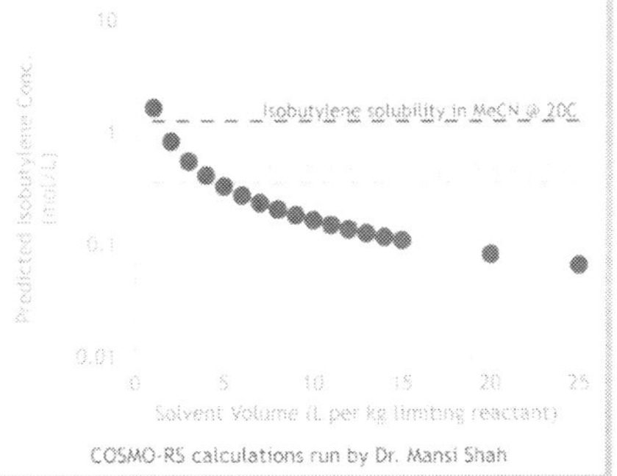

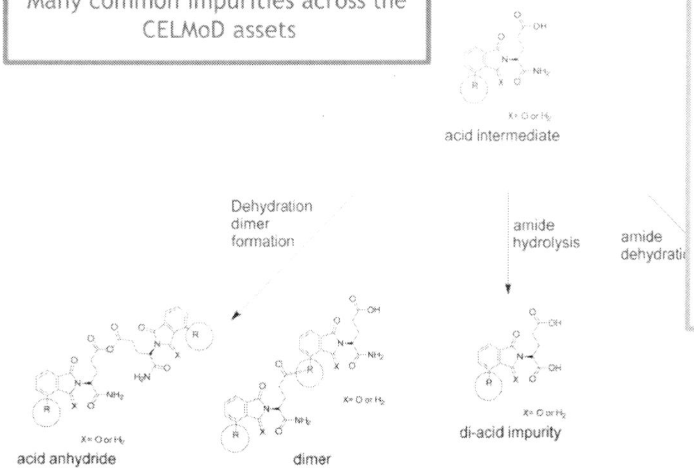

Proposed reaction pathways

Undesired reactions: t-butyl ester deprotection

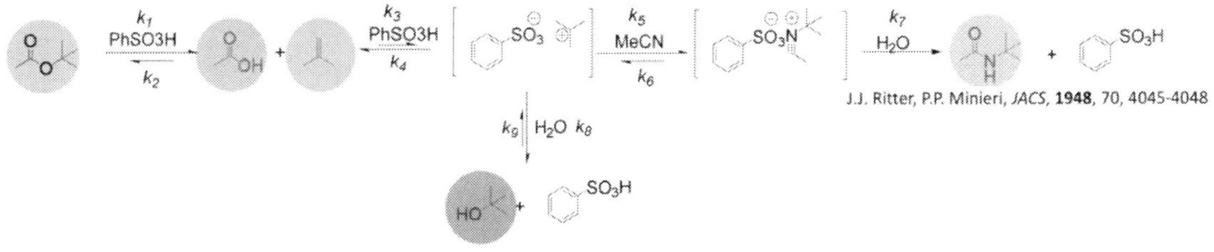

Undesired reactions: t-butyl ester deprotection

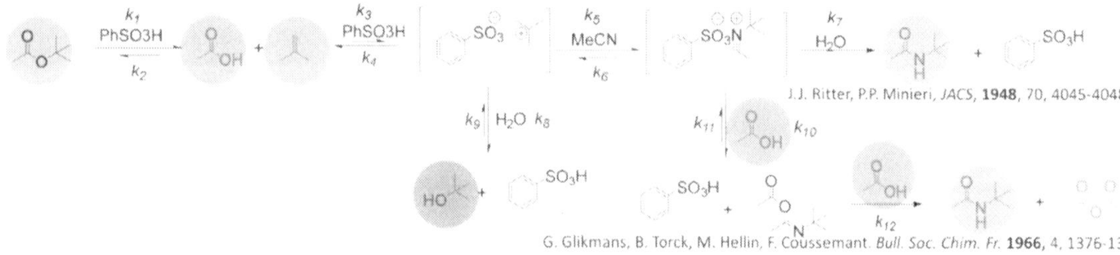

J.J. Ritter, P.P. Minieri, *JACS*, **1948**, 70, 4045-4048

G. Glikmans, B. Torck, M. Hellin, F. Coussemant. *Bull. Soc. Chim. Fr.* **1966**, 4, 1376-1386

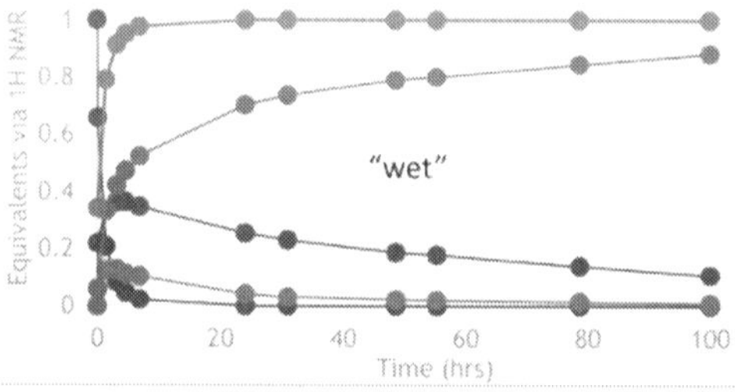

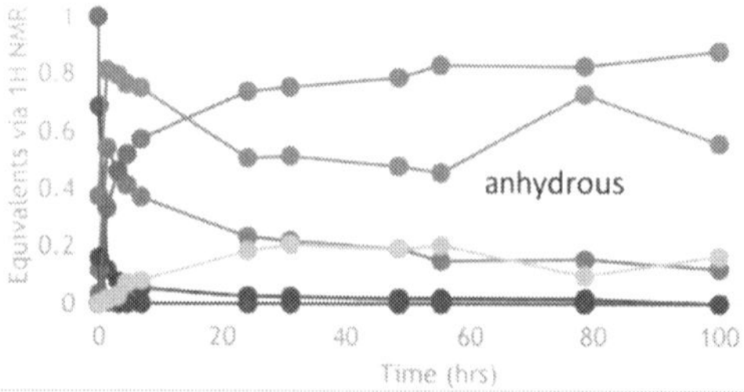

Undesired reactions: cyclization

A series of control experiments with water, isobutylene and benzene sulfonic anhydride were monitored via HPLC

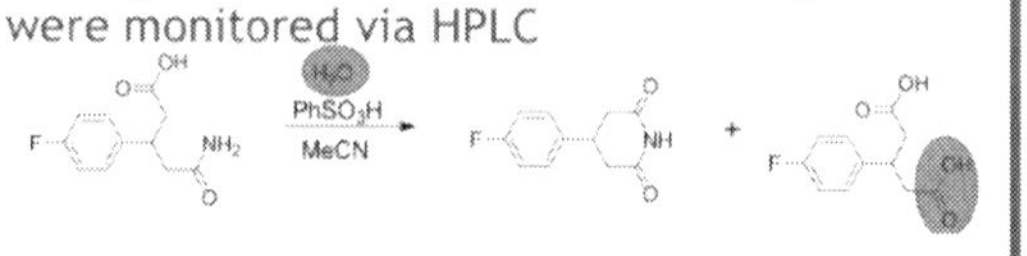

Run control experiments to understand benzene sulfonic anhydride hydrolysis via IR @ 55C

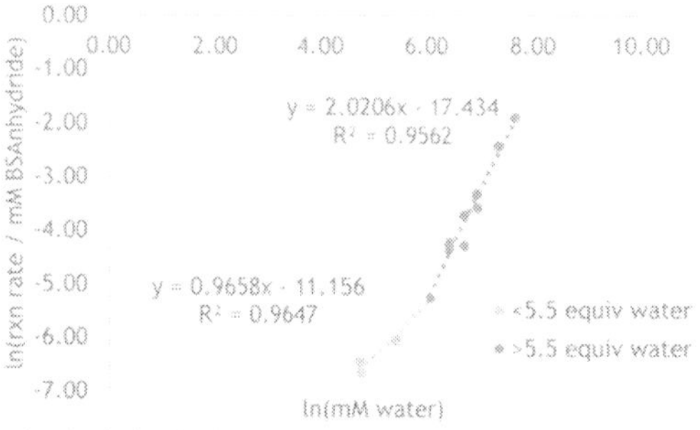

Anhydride hydrolysis is 1st order w.r.t. water at low concentrations but transitions to 2nd order at higher water levels

What about undesired reactions?

Many common impurities across the CELMoD assets, can largely be explained by the interplay between isobutylene and water

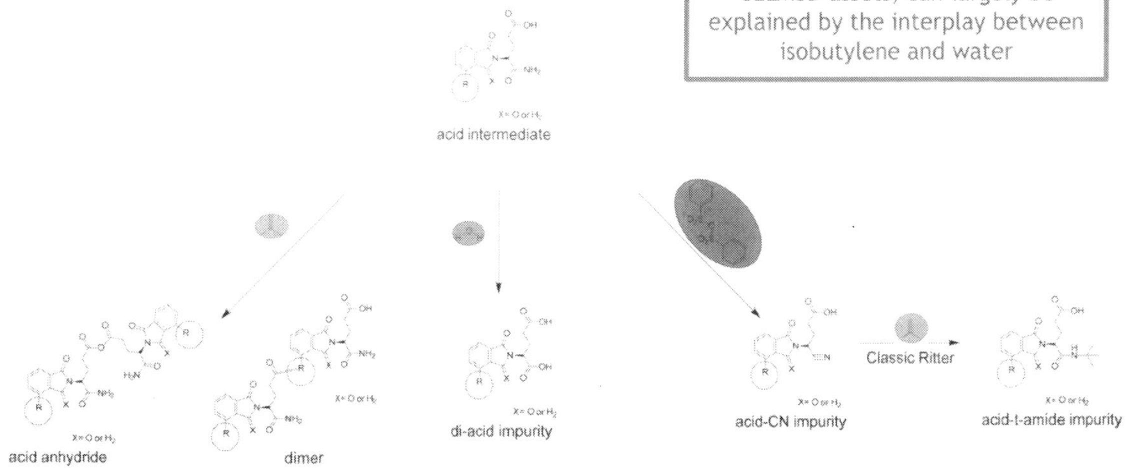

Proposed reaction pathways

Model system key takeaways

t-butyl ester deprotection

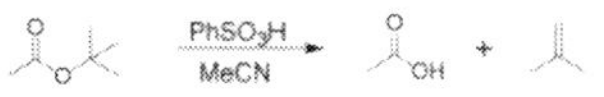

- Acid catalyzed
  - Desired reaction has 2nd order dependence on BSA conc.
- Complex system of side-reactions with isobutylene can lead to a variety of by-products (N-t-butyl acetamide, t-butanol, anhydrides)
- Overall reaction irreversible due to formation of N-t-butyl acetamide

Cyclization

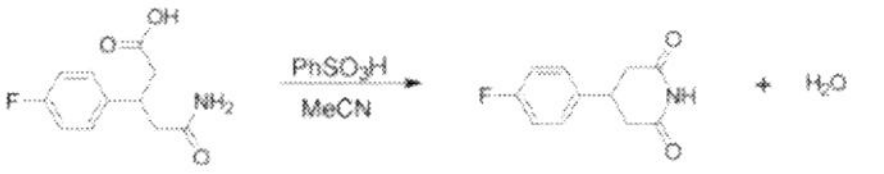

- Acid-catalyzed
  - Desired reaction has 1st order dependence on BSA conc.
- Presence of water leads to di-acid formation
- Presence of excess isobutylene results in high-MW impurities
- Presence of benzene sulfonic anhydride leads to acid-nitrile formation

Key takeaway: isobutylene accumulation leads to impurity formation; robust control strategy will require removal (either chemically or via mass-transfer)

Portfolio Example 1: CELMoD A

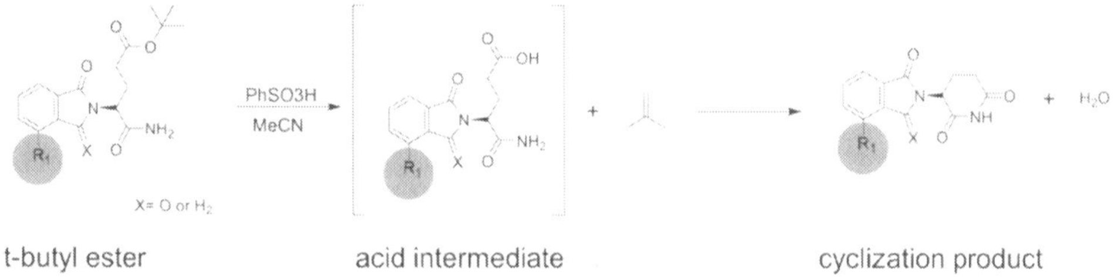

- CELMoD A currently in Phase I clinical trials
- Initial manufacturing campaigns required sub-surface N2 sparging for isobutylene control
  - 2 batches, ~80% yield, >98 LCAP, ~90wt% solids
    - N-t-butyl acetamide present < 10 mol%
- Further development work identified a problematic dimer impurity
- Goal: Optimize process to minimize dimer formation for more consistent reaction performance across scales

Portfolio Example 1: CELMoD A

Typically see up to 2 LCAP of acid-CN, acid-t-bu-amide, and dimer at the end-of-reaction

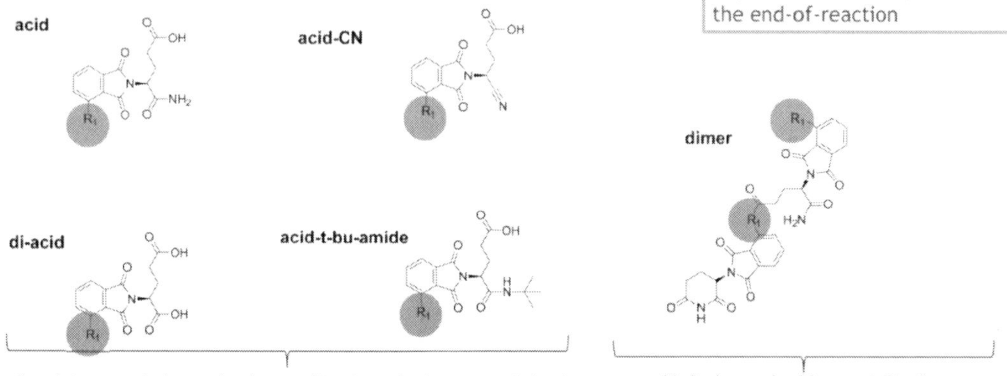

Complete purge during crude salt crystallization and subsequent salt-break

Limited purge in either crystallization

Hypotheses
- Adding water will minimize impurity formation
- Addition of water will be yield impacting
  - Incomplete conversion
  - Formation of di-acid

Portfolio Example 1: CELMoD A Kinetic Model

- Goal: Build kinetic model that captures the dimer and di-acid dynamics
  - Set all reactions* to equilibria and let model adjust K_{eq} and k values
  - Use mechanistically derived equations for desired
  - Where applicable (anhydride hydrolysis) use pre-determined rate constants and Ea to minimize number of equations

t-butyl_ester
BSA
water
MeCN
acid
pdt
Isobutylene
di-acid
acid-CN
acid-t-bu-amide
Ammonium_besylate
t-butyl_cation_complex
N-t-butylacetamide
nitrillium_ion_complex
tBuOH
...

Reactions in Solution

Desired reactions:
- t-butyl_ester + BSA ⇌ acid + Isobutylene + BSA
- acid + BSA ⇌ pdt + water + BSA

- BSA + Isobutylene + MeCN ⇌ nitrillium_ion_complex
- water + BSA + Isobutylene ⇌ tBuOH + BSA
- BSA + nitrillium_ion_complex → bsa-anhydride + N-t-butylacetamide
- nitrillium_ion_complex + water → BSA + N-t-butylacetamide

Impurity reactions:
- bsa-anhydride + water → 2 BSA
- bsa-anhydride + acid → 2 BSA + acid-CN
- acid + BSA + water → di-acid + Ammonium_besylate
- acid + t-butyl_ester + nitrillium_ion_complex → dimer-SM + N-t-butylacetamide + BSA
- dimer-SM + BSA → dimer + Isobutylene + BSA
- 2 acid + nitrillium_ion_complex → dimer + N-t-butylacetamide + BSA
- dimer + water + BSA → 2 acid + BSA

	Volume (L/kg)	Soln. Temp (°C)	Equiv. BSA	Equiv. H2O
ExpA	12	50	4.9	3.3
ExpB	12	55	4.9	3.6
ExpC	12	67	3.5	3.7
ExpD	20	5	8.0	2.1
ExpE	12	20	7.6	0.51
ExpF	11	20	1.0	0.38

*Reactions involving N-t-butylacetamide as the product were modeled irreversible from previous stability studies

Portfolio Example 1: CELMoD A Kinetic Model

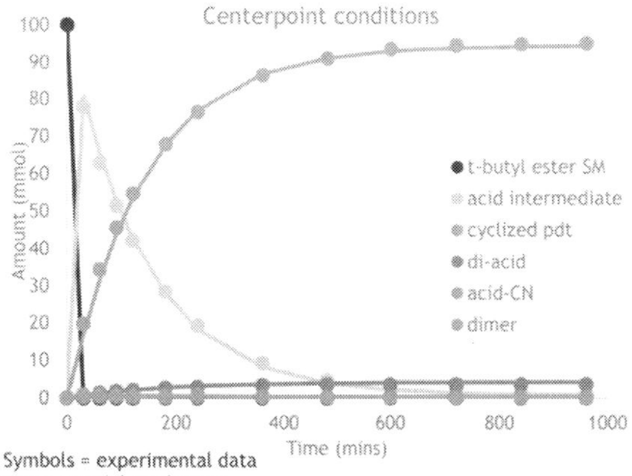

Symbols = experimental data
Lines = model predictions

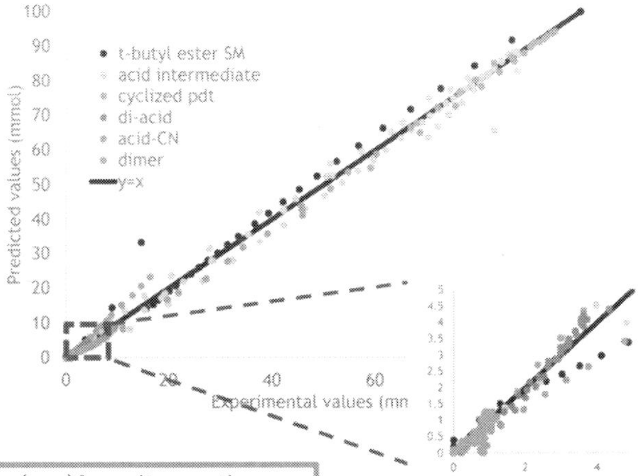

Model captures kinetics of desired and impurity reactions

Portfolio Example 1: CELMoD A Conclusions

- Model enables virtual parameter exploration
  - Dynamics of dimer, di-acid, and acid-CN impurities well-captured
- Model highlights role of water in reducing impurities by:
  - Eliminating dehydration pathways via quenching t-butyl and nitrilium ions
  - Hydrolyzes dimer
- New centerpoint conditions demonstrated on scale

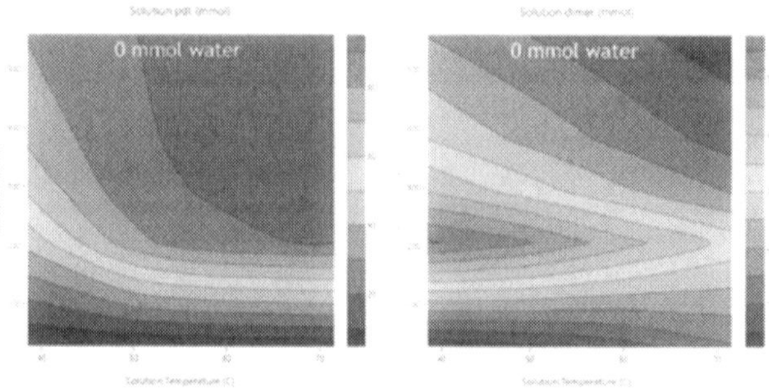

Results from most recent manufacturing campaign (2 kg scale)							
	di-acid	acid	acid-CN	Acid-t-bu-amide	Pdt	Dimer	Yield
End-of-reaction	2.97	1.51	0.11	0.19	94.74	n.d.	
Crude Salt	1.77	n.d.	n.d.	n.d.	98.23	n.d.	86.0
Final API	n.d.	n.d.	n.d.	n.d.	100.00	n.d.	78.6

Portfolio Example 2: CELMoD B

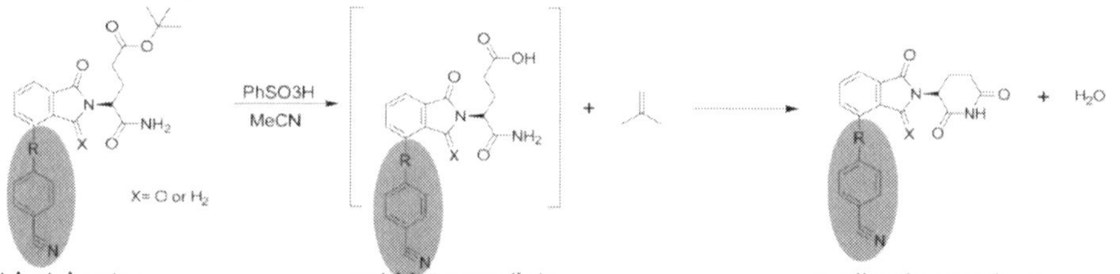

t-butyl ester | acid intermediate | cyclization product

- CELMoD asset currently in Phase II clinical trials
- Some variability observed in manufacturing for some key impurities
- Goal: improve reaction understanding to minimize batch-to-batch variability

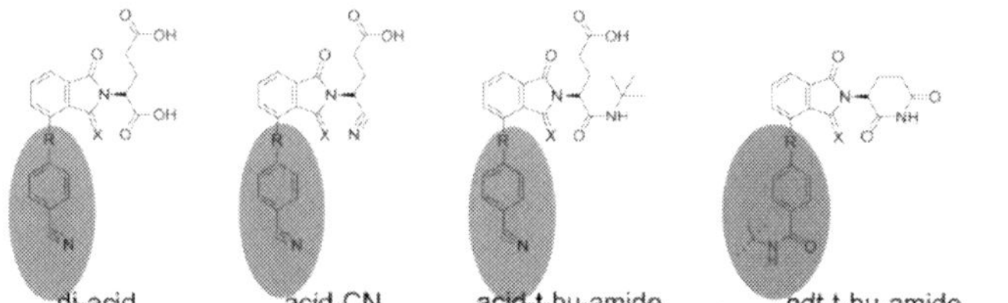

di-acid | acid-CN | acid-t-bu-amide | pdt-t-bu-amide

~90% rejection during crystallization

~50% rejection in crystallization

Portfolio Example 2: CELMoD B

- Addition of water (>1 equiv) minimizes acid-CN and acid-t-bu-amide impurities, substantially increases di-acid impurity, and does not substantially influence pdt-t-bu-amide impurity

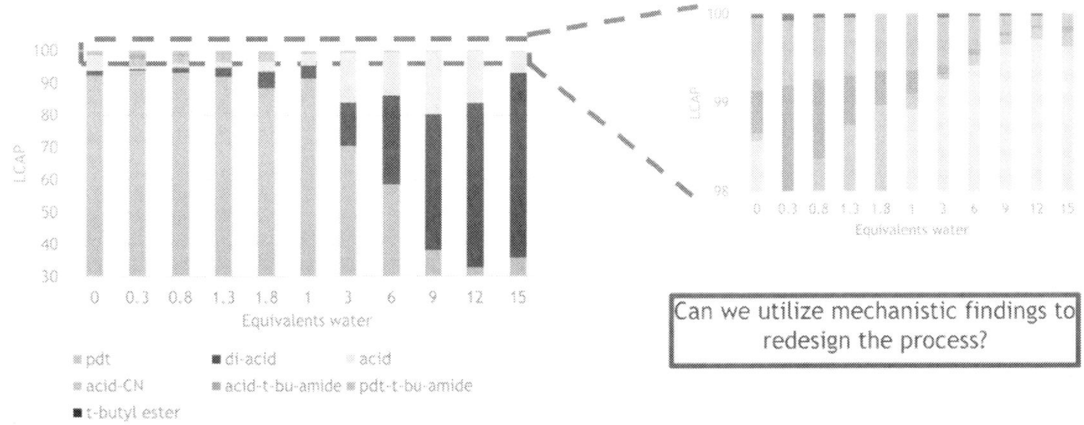

Can we utilize mechanistic findings to redesign the process?

Portfolio Example 2: CELMoD B

- To minimize the formation of pdt-t-amide, we want to prevent accumulation of isobutylene in solution
  - Dilution with MeCN
  - Mass transfer via sub-surface N2 sparging

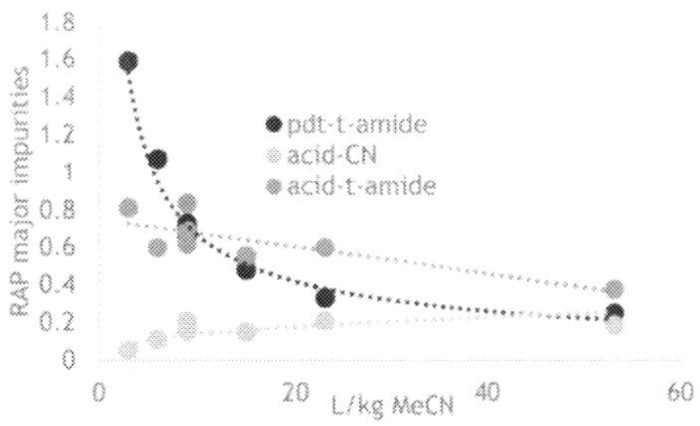

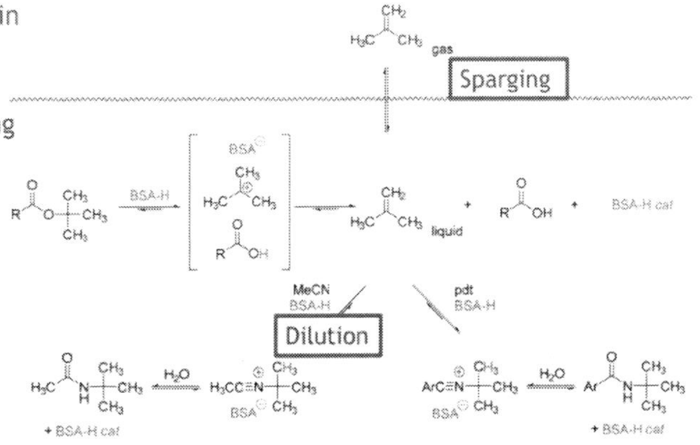

Portfolio Example 2: CELMoD B Kinetic Model

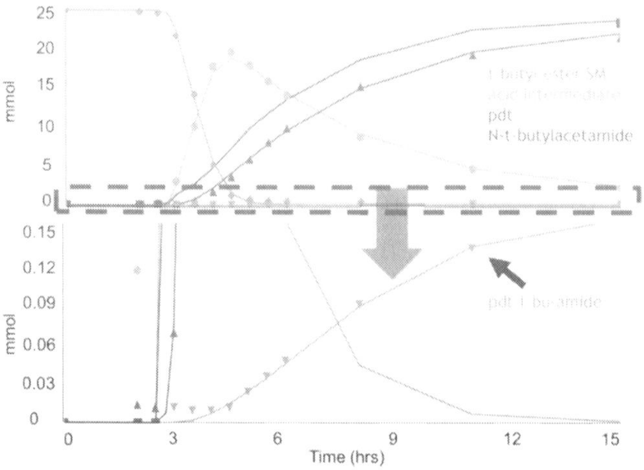

Sensitivity Analysis

Factors	Score
Batch Volume	-0.94
Batch Temperature	0
t-butyl ester	2.25
PhSO3H	0
Water	-0.03
kLa	-0.98

0.0 = nonsensitive
0.5-1.0 = moderate effect
>1.0 = large effect

Higher dilution → Lower levels of key impurity

Higher kLa (via increased T) leads to lower impurity levels

Team successfully ran modified process (70C, 17 L/kg MeCN, sparging) for the validation campaign at an external manufacturer

Conclusions

- Acid-catalyzed transformation developed to install glutaramide as final step in DS process

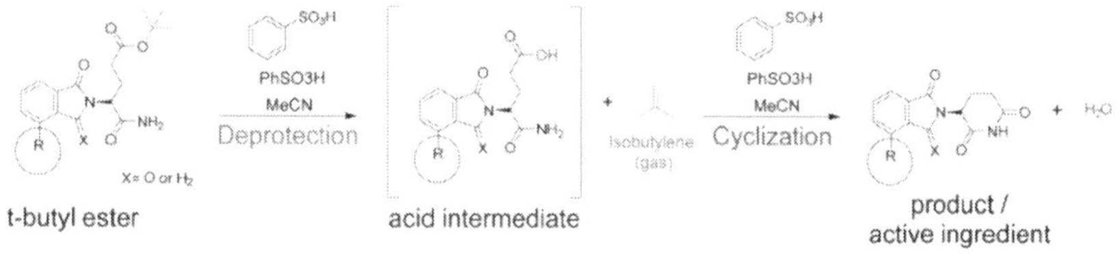

Acknowledgements
CPD leadership team, the Reaction Science Group, and many others who have contributed to the historical knowledge on these programs

- Kinetic models built from mechanistic studies on model systems and data-rich experimentation on CELMoD assets using in-line tools to minimize worker exposure
  - Isobutylene accumulation is the root cause of majority of observed impurities
    - Scale-independent control (water, dilution)
    - Scale-dependent control (sparging)
- Predictive ReactionLab™ models enable parameter space to be explored virtually, limiting worker exposure

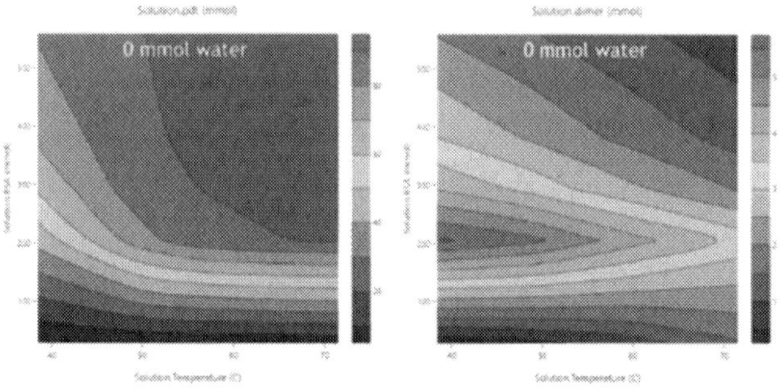

AUTHOR INDEX

AIChE
120 Wall Street, FL 23
New York, NY 10005-4020

ISBN 978-1-7138-6319-9

CURRAN ASSOCIATES INC.
proceedings.com